John Greaves

Treatise on Elementary Hydrostatics

John Greaves

Treatise on Elementary Hydrostatics

ISBN/EAN: 9783744696531

Printed in Europe, USA, Canada, Australia, Japan

Cover: Foto ©ninafisch / pixelio.de

More available books at **www.hansebooks.com**

A TREATISE

ON

ELEMENTARY HYDROSTATICS

BY

JOHN GREAVES, M.A.,

FELLOW AND LECTURER OF CHRIST'S COLLEGE, CAMBRIDGE,
AND FORMERLY ASSISTANT MASTER AT BEDFORD GRAMMAR SCHOOL.

CAMBRIDGE:
AT THE UNIVERSITY PRESS.
1894

[*All rights reserved.*]

Cambridge:
PRINTED BY C. J. CLAY, M.A. AND SONS,
AT THE UNIVERSITY PRESS.

PREFACE.

IN this Book it has been my object to treat the subject as fully as possible without using the Calculus: as, however, it is intended for the use of Students preparing for the First Part of the Mathematical Tripos, and the notation of the Calculus is no longer prohibited in the first four days, alternative proofs have been given where the Calculus enables us either to obtain the results more easily or to express them more concisely.

In Chapter I., after shewing that solids may be classified according to their behaviour under the action of forces, I have deduced the definition of a fluid from the characteristic behaviour of all substances which we recognise as fluids. From this definition the principles of the subject are deduced.

In Chapter III., in addition to the ordinary propositions relating to the distribution of pressure in a homogeneous liquid at rest under gravity, I have given the corresponding ones for a heterogeneous fluid in equilibrium under any system of forces, as well as for certain cases of simple motion. The articles containing the latter results, with certain others intended for a second reading, are marked with an asterisk.

G. E. H.

Chapters VI. and VIII. contain descriptions and diagrams of the most important hydrostatic machines, as well as of the different apparatus for the determination of specific gravities.

In Chapter IX. it is shewn that it follows from certain experiments, that the energy of a material system depends partly on the extent of the surfaces separating the different substances. On the assumption of the existence of this surface-energy, several well-known capillary phenomena are deduced: in some cases alternative proofs depending on the existence of a surface-tension are also given. In connection with this part of the subject I have made considerable use of the Chapter on Capillarity in Prof. Clerk Maxwell's text-book on 'Heat', and of the article contributed by him to the *Encyclopædia Britannica*.

I take this opportunity of thanking my two friends Mr W. B. Allcock, M.A., Fellow and Tutor of Emmanuel College, and Mr H. C. Robson, M.A., Fellow and Lecturer of Sidney Sussex College, for their kindness in revising the proof-sheets, often when very busy with the work of the term, and for their many very valuable suggestions and criticisms. My thanks are also due to my friend Dr Hobson, F.R.S., for his advice on several important points.

I shall be glad to receive any suggestions or corrections.

JOHN GREAVES.

CHRIST'S COLLEGE,
January, 1894.

CONTENTS.

CHAPTER I.

INTRODUCTORY.

	PAGE
Experiments on solids	2
Definition of a fluid	5
Definition of a perfect fluid	6
Distinction between a fluid and a gas	7

CHAPTER II.

PROPERTIES OF FLUIDS.

Definition of pressure at a point	9
Equality of pressure at a point in all directions	11
Definition of Compressibility and Elasticity	14
Transmissibility of liquid pressure	15
Definition of Specific Gravity	16
Specific gravity of mixtures	18
Illustrative Example	20
Examples	20

CHAPTER III.

GENERAL THEOREMS RELATING TO PRESSURE.

	PAGE
Pressure throughout a horizontal plane constant	22
Relation between pressure and depth in a homogeneous liquid	24
Density throughout a horizontal plane constant	26
Surface of separation between two fluids a horizontal plane	27
Surfaces of equal pressure	30
Surfaces of equal pressure and density coincide	34
Fluid revolving about a vertical axis	38
Thrust on a plane area	43
Impulsive pressure	45
Illustrative Examples	48
Examples	56

CHAPTER IV.

CENTRE OF PRESSURE.

Centre of pressure of a parallelogram	64
Centre of pressure of a triangle with horizontal base	65
Centre of pressure of any triangle	68
Centre of pressure of a circle	70
Centre of pressure of any plane figure	72
Illustrative Examples	72
Examples	75

CHAPTER V.

FLOATING BODIES.

Archimedes' Theorem	79
Force and centre of Buoyancy	80
Resultant vertical thrust on surface	81
Resultant horizontal thrust on surface	82

CONTENTS.   ix

	PAGE
Conditions of equilibrium of floating body	86
The Balloon	88
Conditions of equilibrium of body turning about a point	90
Stability of Floating Bodies	91
Surface of Buoyancy	93
Surface of Floatation	94
Determination of metacentre	96
Illustrative Examples	100
Examples	106

CHAPTER VI.

THE DETERMINATION OF SPECIFIC GRAVITY.

Specific Gravity Bottle	116
The U-tube method	118
The inverted U-tube method	119
The Hydrostatic Balance	120
Jolly's Balance	122
The Common Hydrometer	123
Nicholson's Hydrometer	125
Examples	127

CHAPTER VII.

GASES.

Air possesses weight	130
The Barometer	130
Corrections to be applied to a barometric observation	131
The Aneroid Barometer	133
Boyle's Law	134
Perfect Gases	136
The Diving-Bell	137
Charles' Law	138
Absolute temperature	139
Mixture of Gases	140

x   CONTENTS.

	PAGE
Density of the Atmosphere	142
Altitude by Barometric Observations	144
Homogeneous Atmosphere	145
Illustrative Examples	145
Examples	147

CHAPTER VIII.

HYDROSTATIC MACHINES.

The Siphon	153
Torricelli's Theorem	154
The Common Syringe	157
Valves	158
The Common Pump	159
The Lift Pump	160
The Force Pump	161
The Fire Engine	163
Bramah's or the Hydraulic Press	163
Hawksbee's Air-Pump	164
Smeaton's Air-Pump	165
Tate's Air-Pump	166
Sprengel's Air-Pump	167
The Condenser	168
The Barometer Gauge	169
The Siphon Gauge	169
The Condenser Gauge	170
The Compressed-air Manometer	170
Illustrative Example	171
Examples	172

CHAPTER IX.

CAPILLARITY.

Plateau's Experiment	175
Quincke's Experiment	176
General Principle from which Capillary phenomena may be deduced	177

CONTENTS.

	PAGE
Surface-energy	177
Angles between three fluids in contact, constant	178
Angle of contact of liquid with solid, constant	179
Rise of liquid between two parallel plates	180
Rise of liquid inside fine tube	183
Flexible membranes	184
Surface tension	185
Tension at point of membrane same in all directions	186
Tension in cylindrical membrane	187
Lintearia	188
Tension in spherical membrane	188
Work done in stretching membrane	189
Capillary Curve	191
Rise of liquid in fine tube	192
Soap Bubbles	192
Illustrative Example	194
Examples	195
Answers to Examples	197
Index	203

CHAPTER I.

INTRODUCTORY.

1. IN Mechanics material substances are classified according to their behaviour under the action of forces. In Statics and in the Dynamics of a Rigid Body we investigate the effect of a force on an ideal body, a *perfectly rigid* one, i.e. one that moves as a whole only, never undergoing any alteration in either shape or size. We know however from actual experience that very many bodies when acted on by force are perceptibly altered in shape, size, or in both. In some bodies unless the forces are very large, the alterations may be imperceptible, but it is very difficult to believe that when different portions of the body mutually act on one another, the body does not yield in any way.

Let us consider the mutual actions between the two portions of a substance lying immediately on either side of a small plane drawn anywhere within the substance. The two actions form a **stress**: they may be resolved at right angles to the plane and along it. The two forces perpendicular to the plane will be either *pulls* or *thrusts*, according as they tend to keep the two portions together, or to keep them apart, and together constitute a **normal stress**: the two along the plane form a **tangential** or shearing stress. A normal stress is generally accompanied by a *compression* or an *extension* according as it consists of thrusts or pulls. A shearing stress is generally accompanied by the matter on one side the plane sliding over the other. These deformations are termed **strains**; and the latter is termed a **shearing strain** or simply a **shear**. The deformation produced by a pair of scissors used in the ordinary way is an illustration of a shear.

2. The following experiments illustrate the different ways in which different substances behave when they have been deformed and the stress producing the deformation is removed.

(*a*) Stretch a steel wire by hanging a weight at one end and fixing the other end: it will be found that if the weight be not too large, the wire will resume its original length when the weight is removed.

(*b*) Press the two prongs of a tuning-fork towards one another; when released they will spring back into their original position.

(*c*) Push an inverted tumbler containing air below the surface of some water so as to compress the air; when the tumbler is removed the air will resume its original bulk.

(*d*) Press your finger against a piece of wet clay; when it is removed a perceptible dent is left on the surface of the clay.

We notice then a great difference between the behaviour of the steel and air in experiments (*a*), (*b*) and (*c*) and that of the clay in experiment (*d*). When the particular stress in question is removed, in the case of the steel or air, the corresponding strain disappears, whereas in the case of the clay the deformation is permanent. We express these facts by saying that for the stresses in question steel and air are **perfectly elastic,** whereas the clay is not, except possibly when the stresses are very small. On account of this property clay is termed **plastic.** It should be noticed that the strains produced in the experiments are different one from another, and that it does not follow that because a substance is elastic for one sort of strain it is so for another: thus air, which is perfectly elastic for a change in bulk, is not so for a change in shape: we must therefore distinguish between **Elasticity of bulk** and **Elasticity of shape.** Nearly all substances possess the former property.

3. Let us now make the following experiments. Increase the weight suspended from the steel wire in experiment (a) very considerably, and it will be found that the strain produced in the wire does not disappear when the weight is taken off, but is permanent. Again, if a slight blow be given to a glass tumbler, the tumbler rings for a short time, indicating that it is vibrating and consequently undergoing a change of shape, and then resumes its old shape. If however a violent blow be given to the tumbler, it breaks.

From these experiments we see that a body may be perfectly elastic within certain limits but not beyond them. If when the limits of perfect elasticity are exceeded the body breaks like glass, it is termed **brittle.** If on the other hand it bends and remains bent rather than breaks, like wrought iron, it is termed **tough.**

In the experiments we have considered so far the

strain in each case has been produced practically at once and has not increased when the stress occasioning it has acted for a longer time. If on the other hand a lump of pitch be placed on a table, it is found that the longer the time it is left the more it flattens itself out. When the strain produced by a particular stress increases with the time during which the stress acts, the substance is termed **viscous**. The more slowly the strain is produced the greater is the viscosity.

Plastic bodies sometimes yield more readily to one change of shape than to another; thus some like copper are easily drawn out into wire and are termed **ductile**; others like gold are easily hammered out into sheets, and are termed **malleable**; others again like moist clay or wax can be easily moulded into any shape and are termed **soft**.

4. We have seen then that substances are termed elastic, viscous, &c. according to the different ways in which they behave under the action of forces. What is the essential difference between the behaviour of a solid and that of a fluid? We are all aware that a mass of any substance which we recognise as a solid if placed on a flat table quickly settles down into a position of equilibrium.

Let ABC be this mass in its position of equilibrium.

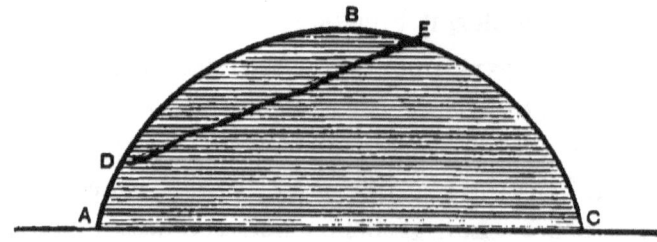

Let DE be a plane inclined to the horizon, dividing the solid into two portions.

The weight of the upper portion would cause it to slide down the plane, if there were no shearing stress to prevent such motion. We infer then from this that every solid can resist a shearing stress without giving way to an indefinite extent. If now a considerable mass of water, glycerine, pitch or any substance which we recognise as a liquid be placed on a flat table, it will spread itself out continuously, quickly in the case of the water, and very slowly in the case of the pitch, and will never be found in equilibrium except with the surface exposed to the air horizontal. As the above liquids can be supported to any depth in a vessel, they must be capable of exerting normal stresses; but as equilibrium is impossible when shearing stresses however small are required to maintain it, we infer that the above substances cannot exert shearing stresses when they are in equilibrium.

We cannot perform an experiment similar to the last with a gas, but we are aware of the experimental fact that if a gas be allowed access to any closed vessel, it will fill every nook of that vessel, even though a small shearing stress would prevent this. This points to the conclusion that no shearing stress is exerted in a gas at rest.

We are thus led up to the following definition of a fluid:—

A Fluid is a substance which will yield to any continued shearing stress however small: in other words, **when a fluid is in equilibrium, the stress across any plane in it is entirely normal to that plane.**

It should be observed that a fluid will yield *in time* to the *smallest* shearing stress. It may however resist a considerable shearing stress, if applied for only a short time: thus, a sudden blow may produce no perceptible impression on a lump of pitch, which we regard therefore as a *hard liquid*. A similar blow will, on the other hand, produce a visible dent in a lump of putty, though the putty will resist a small shearing stress for an indefinite period of time. It is therefore a *soft solid*.

5. If three bottles of the same shape and size be filled respectively with water, glycerine and treacle, and then held upside down, it will be found that the water escapes first and the treacle last. Again, if these three liquids be each placed in a separate tumbler and set rotating by means of a spoon, it will be found that the treacle comes to rest first and the water last. If there were no friction, or shearing stress, the bottles would empty themselves in the same time if *in vacuo*, and the liquids in the tumblers would go on rotating indefinitely. We infer then that when liquids are *in motion*, the stresses exerted are not entirely normal. The same is true of gases, since when set rotating in a closed vessel they gradually come to rest. It is obvious, however, from the experiments just described that the shearing stresses in the glycerine are less than those in the treacle, but greater than those in the water. This fact is expressed by saying that of the three, treacle is the most viscous, water the least. We are thus led on to the idea of a fluid in which there is no shearing stress even when it is in motion, though we have no experience of such a one. Such an ideal substance is termed a **perfect fluid.**

DEF. **A perfect fluid is one in which no shearing stress is exerted, whether the fluid be at rest or in motion.**

6. We have seen that the term *fluid* includes both *liquids* and *gases*: how then do we distinguish between them? If a quantity of a gas be enclosed in a vessel, and the vessel be enlarged to any extent, it is found that the gas will still occupy the whole vessel: this is certainly not true of a liquid.

DEF. **A Gas is a fluid, a given portion of which can be made to expand indefinitely by increasing sufficiently the space to which it has access.**

A Liquid is a fluid, the volume of a given portion of which never exceeds a definite amount, no matter to how large a space it has access, or how small the pressure to which it is subjected.

It is an experimental fact that when the volume of a liquid in a closed vessel is less than that of the vessel, the liquid evaporates until the part of the vessel not occupied by the liquid is filled with vapour. It might then be objected to the above definition of a liquid that its vapour can expand indefinitely. This is true, but does not affect the definition, because it is not the liquid as such which expands, but the vapour into which a portion of it is converted.

It has been proved experimentally that all gases can be converted into liquids by lowering their temperature and increasing the pressure on them sufficiently. When however the temperature of a gas is above a certain point no increase of pressure alone will bring about a condensation, whereas when the temperature is below this point, a sufficient increase of pressure produces condensation. This temperature which varies for different gases is called the *critical temperature*. When the temperature of a gas is below its critical point, the gas is termed a **vapour,** and when above, a **permanent gas** or simply a **gas.**

Another property characteristic of a gas is that when it is in the presence of another gas there is no definite surface of separation between them; whereas, when a liquid is in the presence of a gas or of its own vapour, there is a definite surface of separation.

Dr Andrews' experiments on the liquefaction of Carbon Dioxide have shewn that the properties of a liquid and its vapour may approach so near to one another that it is impossible to say in which state the substance is.

7. In Art. 4 we were led up to the characteristic property of a fluid by considering the behaviour of liquids and gases in certain circumstances: now it may have suggested itself to the student that *small* quantities of liquids under similar circumstances behave differently. Thus a drop of mercury placed on a flat piece of glass does not spread itself out flat but rolls itself up into a ball: and similarly drops of dew are small spheres of water collected on the leaves of plants. We are not justified then in asserting that in the case of small quantities of liquids at rest there are no shearing stresses. These phenomena and others, which are due to Capillarity, point to the conclusion that small stresses, other than normal thrusts, act among the particles on the surface of a liquid. When we are considering small quantities of liquids, either in the form of drops or in fine tubes, the effect of the surface forces is very appreciable, but when the quantity of liquid is large, the effect is so small, that we shall not take into account the capillary forces.

CHAPTER II.

Properties of Fluids.

8. When any plane area is acted on by a thrust, the *pressure* on the area is the intensity of the thrust, and may be *uniform* or *varying*.

Def. **Uniform Pressure.** *The pressure on a plane area is uniform when the thrust on any portion of it is proportional to the area of that portion, and the pressure is measured by the thrust on a unit area.*

When the thrust on any portion is *not* proportional to its area, the pressure is *varying*.

Def. **Mean Pressure.** *The mean pressure on a plane area is the uniform pressure on it which will give the same resultant thrust as the actual one.* Hence, if A be the area, and P be the resultant thrust on it, the mean pressure is P/A.

Def. **The pressure at any point of an area is the limit of the mean pressure on an indefinitely small area enclosing the point.**

The idea of a varying pressure may become clearer to the student, if he observes the analogy it bears to a varying velocity. When we assert that the velocity of a body at a particular instant is 30 feet per second, we do not mean that it will of necessity travel 30 feet in the next second, because its velocity may change in the meantime, but that it will travel at *that rate* for an indefinitely small interval of time containing the instant in question. If a body travel 30 millionths of a foot in one millionth of a second, its mean velocity during that portion of a second is 30 ft. per sec. Similarly if the thrust on 1/1000000 square inches is 3/1000000 lbs., the mean pressure on that small area is 3 lbs. per sq. in. This will approximately be the pressure at any point of the small area, but we cannot assert that the thrust on a square inch is 3 lbs.

9. If the force on a plane area be normal, but a *pull* instead of a *thrust*, we must substitute **tension** for **pressure** in the above, and we shall obtain the meaning of the tension at any point when uniform and when varying.

If the force be **tangential** we can define in a similar way the **tangential force per unit area** at any point.

It is easily shewn that pressure and the tangential force per unit area are of one dimension in force and -2 in length, i.e. of 1 dimension in mass, -1 in length, and -2 in time.

Ex. 1. If the thrust on an area of 4 sq. yards be 1000 lbs. weight, and the pressure over it be uniform, find its magnitude in ft. lb. sec. units.

Ex. 2. A solid rectangular parallelopiped, weight 1 kilogramme, and edges of length 12 cm. 6 cm. and 2 cm. respectively, rests on a horizontal table. If the pressure on the table be uniform, find its magnitude when the different faces are respectively in contact with the table, in C.G.S. units.

10. If we consider the normal stresses across different planes drawn through a point in a solid, we shall find that generally they are different in magnitude. For instance if we consider a particular brick in a wall, the thrust per unit area on a horizontal face, due to the weight above it, is probably different from that on a vertical one, due perhaps to the thrust of a buttress or an arch. The pressure then at any point of a substance will in general depend on the direction of the plane across which we are estimating the stress. We shall prove however that if there is no tangential stress across any plane, as in the case of a fluid at rest, the pressure at any point is the same for all planes through the point.

11. PROP. The pressure at any point of a fluid at rest is the same for all directions.

(a) *For a single external force, viz. gravity.*

Let O be any point in the fluid. From O draw OA, OB, OC mutually at right angles, and join AB, BC, CA.

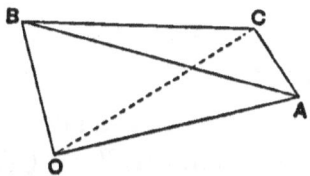

Let p_A, p be the mean pressures across the faces BOC, ABC, and let ρ be the density (Art. 16) of the fluid within the tetrahedron $ABCO$.

The forces acting on the fluid within the tetrahedron are

(1) its weight $\tfrac{1}{6}g\rho \cdot OA \cdot OB \cdot OC$ in direction making some angle θ, with OA,

(2) the thrust on the face OBC, $\tfrac{1}{2}OB \cdot OC \cdot p_A$ along OA,

(3) the thrust on the face ABC, $p \cdot \text{area } ABC$, at right angles to ABC,

(4) the thrusts on the faces AOB, AOC.

Since the triangle BOC is the projection of the triangle ABC on the plane BOC,
$$\triangle BOC = \triangle ABC \cos \phi,$$

where ϕ is the angle between ABC and BOC, i.e. between OA and the perpendicular to ABC.

Since the fluid tetrahedron is in equilibrium, resolving along OA, we have

$$\tfrac{1}{2} OB \cdot OC \cdot p_A - \triangle ABC \cdot p \cos \phi$$
$$+ \tfrac{1}{6} g\rho \cdot OA \cdot OB \cdot OC \cos \theta = 0,$$
$$\therefore \tfrac{1}{2} OB \cdot OC \cdot p_A - \tfrac{1}{2} OB \cdot OC \cdot p + \tfrac{1}{6} g\rho \cdot OA \cdot OB \cdot OC \cos \theta = 0;$$
$$\therefore p_A - p + \tfrac{1}{3} g\rho \cdot OC \cos \theta = 0.$$

When OA, OB, OC are taken indefinitely small, this equation reduces to $p_A = p$, and the mean pressures on the faces become the pressures at O in the corresponding directions.

Hence the pressure at O along OA is equal to that perpendicular to ABC.

In a similar way we can shew that the latter is equal to the pressure along OB or OC, and by varying the magnitude of OA, OB and OC, we can shew that this holds whatever the direction of the plane ABC.

NOTE. We have assumed in the above that the density is uniform; if it is not, we may regard ρ as the mean density of the tetrahedron and the proof still holds.

(β) *For any system of external forces.*

In this case we shall have instead of the average weight, the average resultant external force, and instead of the vertical, we shall have the direction of the average resultant external force. Proceeding as before we obtain the same result.

It should be observed that the reason the resultant external force in (β) and the weight in (α) of the indefinitely small tetrahedron disappear is that as they depend on the volume of the tetrahedron, they are of the third order of small quantities, whereas the thrusts, depending on the areas of the faces, are of the second order only.

PROPERTIES OF FLUIDS. 13

12. We have proved the last theorem for a fluid *at rest*: it can however be extended to the case of a fluid *in motion*, provided the motion is such that there is no tendency of one portion to slide over another, so that no friction or tangential stress is exerted. As illustrations of such motion we may take that of the whole fluid in a given direction with uniform acceleration, or the rotation of the whole fluid as if solid about a fixed vertical axis with uniform angular velocity. For considering an indefinitely small element of the fluid in the shape of a tetrahedron as before, we know that the thrusts on the faces and the resultant external force give the element its resultant acceleration: therefore by Newton's Second Law of Motion, the resolved parts of the thrusts along an edge of the tetrahedron + the resolved part of the resultant external force = the mass of the element × the acceleration along the edge.

But the resultant force and the mass both depend on the volume of the element and are therefore small quantities of the third order, while the thrusts depending on the areas of the faces are of the second order. Hence it follows as in Art. 11, that the pressures are the same in all directions.

It is obvious from the above, that in a *perfect* fluid, which can exert no tangential stress whether it is at rest or in motion, the pressure at a point is *always* the same in all directions.

13. PROP. *The resultant thrust on an element of fluid at rest whose linear dimensions are all indefinitely small is opposite to the resultant external force.*

For considering the equilibrium of the element the

only forces acting on it are the resultant thrust and the resultant of the external forces—they must therefore be equal and opposite. Also since the latter is a small quantity of the third order, the former is one also.

In the case of a *perfect* fluid in motion, and of any fluid for the motion described in Art. 12, the resultant thrust and the resultant external force on an element have a resultant equal to the mass into the acceleration of the element, in the direction of the acceleration.

14. It is an experimental fact that all fluids are more or less compressible: for most liquids, however, the alteration in volume produced by any increase in pressure that is not very great, is so small that it is neglected, and the liquid is for practical purposes *incompressible*.

DEF. *When the volume of any substance is reduced, the consequent* **compression** *is measured by the ratio of the reduction in volume to the original volume.*

DEF. **The compressibility** *of a fluid is measured by the limit of the ratio of the compression produced to the increase of pressure producing it, when this increase is indefinitely small.* Thus, if p be the original pressure, p' the new pressure, v the original volume, and v' the new volume, the compressibility

$$= \frac{v-v'}{v} \Big/ (p'-p), \text{ or } -\frac{1}{v} \cdot \frac{dv}{dp}.$$

DEF. **Elasticity.** *The elasticity of a fluid is measured by the limit of the ratio of an indefinitely small increase in pressure to the compression produced by it.* Thus, if p be the original pressure, p' the new pressure, v the original volume, and v' the new volume, the elasticity

$$= (p'-p) \Big/ \frac{v-v'}{v} \text{ or } -v\frac{dp}{dv}.$$

PROPERTIES OF FLUIDS.

Ex. 1. If the compression in water produced by an increase of pressure of 14·5 lbs. wt. per sq. inch be ·00005, find what diminution of volume, a pressure of 1000 lbs. per sq. in. will produce in 100 cubic feet. What is the elasticity of water in ft. lb. sec. units ?

Ex. 2. A cylinder, axis vertical, and 2 sq. in. in section contains water to a depth of 10 ft. A close fitting piston weighing 600 lbs. is placed on the water so as to be supported by it, find what alteration in the depth of the water it will produce.

15. Prop. Transmissibility of liquid pressure. *An increase of pressure at any point of an incompressible liquid at rest in a closed vessel under the action of any system of external forces is transmitted without change to every other point.*

Let A, B be any two points in the liquid.

(a) *When the straight line AB lies entirely in the liquid.*

Construct an indefinitely thin cylinder having AB for axis.

The liquid contained within the cylinder is kept in equilibrium by

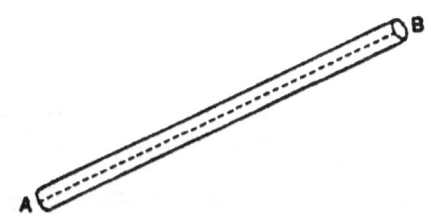

(1) the thrusts on the ends at A and B,
(2) the thrusts on the curved surface, which must be at right angles to AB,
(3) the resultant external force.

Resolving along AB, we see that the difference of the thrusts on the ends at A and B is equal to the resolved part along AB of the external force, i.e. is constant.

Hence the pressure at B always differs by a constant amount from the pressure at A.

(β) *When the straight line AB is not entirely in the liquid.*

Join AB by a series of straight lines AP, PQ, QR, RS, SB lying entirely in the liquid.

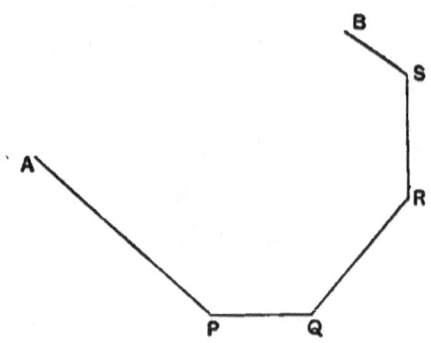

Then the pressure at A — that at $P =$ a constant,
.... $P -$ .... $Q =$ ....,
.... $Q -$ .... $R =$ ....,
.... $R -$ .... $S =$ ....,
.... $S -$ .... $B =$ ....;

∴ the pressure at A — the pressure at B is constant.

Specific Gravity.

16. DEF. *The specific gravity of any substance is the ratio of the weight of any volume of that substance to the weight of an equal volume of some standard substance.*

PROPERTIES OF FLUIDS.

The substance usually adopted as the standard is pure distilled water at a temperature of 4° C.

In text-books on Mechanics the *density* of a substance has been defined as the *mass contained in a unit volume;* hence it follows that the *specific gravity of a substance is the ratio of its density to that of the standard substance.*

For the ratio between the *weights* of equal volumes of the substance and the standard substance is equal to the ratio between the corresponding *masses*, and this again to the ratio between the densities.

Density is shewn in works on Mechanics to be of one dimension in mass, and −3 in length: the density of a given substance will therefore vary with the units of mass and length. On the other hand specific gravity being a ratio merely, depends only on the standard substance and not on the fundamental units.

In the c. g. s. system of units, the unit volume is the cubic centimetre, and the unit mass is that of a cubic centimetre of water, so that the density of water is 1, the same as its specific gravity, if water be the standard substance. In the British system of units, the unit volume is a cubic foot, and the unit mass the Imperial Pound, so that as the mass of a cubic foot of water is about 1000/16 lbs., the density of water is 1000/16 nearly.

17. PROP. *The weight of a body varies as its volume and its specific gravity conjointly.*

Let V be the volume of the body, W its weight, and S its specific gravity.

Let $w =$ the weight of a unit volume of the standard substance.

∴ $wS =$ the weight of a unit volume of the given body.

∴ $wVS =$ the weight of a volume V of the body;

∴ $W = wVS$;

∴ $W \propto VS$.

G. E. H.

18   PROPERTIES OF FLUIDS.

If $w = 1$, i.e. if the unit weight be the weight of a unit volume of the standard substance,
$$W = VS.$$

18. Prop. *To find the specific gravity of a mixture of substances of given volumes and specific gravities.*

Let V_1, V_2, V_3 ... be the volumes of the different substances, S_1, S_2, S_3 ... their respective specific gravities.

Let σ be the specific gravity of the mixture. Then if the volume of the mixture is the sum of the volumes of the components, i.e. $V_1 + V_2 + V_3 + ...$; its weight must be $\sigma(V_1 + V_2 + V_3 + ...)w$, (Art. 17).

But the weight of the mixture = the sum of the weights of the component parts
$$= V_1 S_1 w + V_2 S_2 w + V_3 S_3 w + ...;$$
$$\therefore \sigma(V_1 + V_2 + ...)w = (V_1 S_1 + V_2 S_2 + ...)w;$$
$$\therefore \sigma = \frac{V_1 S_1 + V_2 S_2 + ...}{V_1 + V_2 + ...}.$$

19. Prop. *To find the specific gravity of a mixture of substances of given weights and specific gravities.*

Let W_1, W_2, ... be the weights of the different substances, S_1, S_2, ... their respective specific gravities.

$\therefore$ their volumes are $\dfrac{W_1}{S_1 w}$, $\dfrac{W_2}{S_2 w}$, ... (Art. 17).

Then if the volume of the mixture is equal to the sum of the volumes of the component parts, the volume of the mixture is
$$\left(\frac{W_1}{S_1} + \frac{W_2}{S_2} + ...\right)\frac{1}{w};$$

PROPERTIES OF FLUIDS.

∴ if σ be its specific gravity, its weight is

$$\left(\frac{W_1}{S_1} + \frac{W_2}{S_2} + \ldots\right)\sigma;$$

$$\therefore \left(\frac{W_1}{S_1} + \frac{W_2}{S_2} + \ldots\right)\sigma = W_1 + W_2 + \ldots;$$

$$\therefore \sigma = \frac{W_1 + W_2 + \ldots}{\dfrac{W_1}{S_1} + \dfrac{W_2}{S_2} + \ldots}.$$

NOTE. It has been assumed in this article and the preceding that the volume of the mixture is the sum of the volumes of the component parts; the results therefore will not hold for cases where there is an expansion or shrinkage in volume, as for instance when a salt is dissolved in water.

EXAMPLES.

1. The specific gravity of sea-water being 1·03, find how much fresh water must be added to a gallon of it to reduce its specific gravity to 1·01.

2. Three pints of a liquid whose specific gravity is ·8 are mixed with five pints of another liquid whose specific gravity is 1·04. Find the specific gravity of the mixture if there is a contraction of 5 per cent. on the joint volume.

3. Two volumes of specific gravities s, s', and of volume v, v', having been mixed, the specific gravity of the mixture is found to be σ. Find the volume of the mixture. [M. T., 1864.]

4. Shew that the specific gravity of a mixture of given substances is greater when equal volumes are taken than when equal weights are taken.

5. The specific gravity of a mixture of equal volumes of two substances is S_1 and that of a mixture of equal weights of the same substances is S_2, determine the specific gravities of the two substances.

6. A substance whose specific gravity is ·7 is dissolved in ten times its own weight of water, and the specific gravity of the solution is 1·01, find by how much the total volume is reduced.

ILLUSTRATIVE EXAMPLE.

If the units of mass, length, and time be respectively a lbs., b ft., and c secs., compare the standards in the formulæ $W = g\rho V$, *and* $W = SV$.

What relation must exist among the fundamental units in order that the standards may be the same? [S. John's Coll., 1887.]

Since the three fundamental units are given, the unit of weight is determined, and the value of g, which is clearly $32c^2/b$, if 32 be taken as its value when a foot and a second are units.

In the formula $W = g\rho V$, it is assumed that the standard substance is of unit density, and therefore putting $V=1$, and $\rho=1$, we obtain that the weight of a unit volume is g or $32c^2/b$.

In the formula $W = SV$, it is assumed that the standard substance is of unit specific gravity, and therefore putting $V=1$, and $S=1$, we obtain that the weight of a unit volume is 1.

Hence the density of the standard substance in the first formula is g times that in the second, where $g = 32c^2/b$.

The condition that the standard may be the same for each formula is that $g=1$, or $32c^2 = b$.

EXAMPLES. CHAPTER II.

1. If the specific gravity of water be unity, and a grain be taken as the unit of weight, what must be the unit of volume in order that it may be always true that the weight of any body is equal to the product of the volume and the specific gravity?
[M. T., 1866.]

2. If in the equation $W = g\rho V$, the number of seconds in the unit of time be equal to the number of feet in the unit of length, the unit of weight be 750 lbs. and a cubic foot of the standard substance weigh 13500 oz., find the unit of time. [M. T., 1872.]

3. Taking the pressure of the atmosphere as equal to $14\frac{1}{2}$ lbs. wt. per square inch, find its value in dynes per centimetre, assuming that a gramme is ·0022 of a pound, and that a metre is 39 inches.
[Pet. 1888.]

4. Prove that if the elasticity of a fluid is equal to the pressure, the pressure is inversely proportional to the volume. [M. T., 1875.]

PROPERTIES OF FLUIDS.

5. If the unit of pressure be that of the atmosphere, the unit of angular velocity that of the earth's rotation, the unit angle a radian, and the unit of acceleration that of gravity, compare roughly the density of the standard substance with that of water.

[M. T., 1891.]

6. If the linear dimensions of a fluid medium at rest under parallel forces uniformly distributed throughout it be varied uniformly in the ratio $1 : n$, shew that the pressure at any point is varied in the ratio $n^2 : 1$; and that, if A, B, C be three specified elements of the fluid, the moment of the thrust on the plane ABC, about the line AB, is varied in the ratio $1 : n$. [M. T., 1878.]

7. Shew that if in a certain substance the stress at every point across three planes not parallel to the same line be normal and the same for all three, the substance is a fluid.

[Christ's Coll., 1891.]

8. If the stress across a series of parallel planes in a substance at rest be uniform and tangential, prove that at every point there is another plane across which the stress is also tangential and the same per unit area. [M. T., 1877.]

9. Prove that if in a certain medium there is no shearing stress across any plane parallel to either of two given planes, there is none across one perpendicular to both planes. [Christ's Coll., 1891.]

CHAPTER III.

GENERAL THEOREMS RELATING TO PRESSURE.

20. Prop. *If a fluid, homogeneous or heterogeneous, be at rest under the action of gravity, the pressures at any two points in a horizontal plane are the same.*

(It is assumed that we can join the two points by a horizontal line, curved or otherwise, lying entirely in the fluid.)

Let A, B be the two points.

We shall consider two cases:

(α) When the straight line AB lies entirely in the fluid.

About AB as axis construct a cylinder of indefinitely small section.

Consider the equilibrium of the fluid contained within this cylinder.

The forces acting on this fluid are

(i) its weight, vertical, and therefore perpendicular to AB,

(ii) the thrusts on the curved surface, everywhere perpendicular to the surface, and therefore perpendicular to AB,

(iii) the thrusts on the ends at A, B, along AB.

∴ the thrust on the end at A = that on the end at B: and the ends are equal in area.

∴ the pressure at A = the pressure at B.

(β) When the straight line AB does *not* lie entirely in the fluid.

We can in this case join AB by a broken horizontal line $APQRSB$, the several parts AP, PQ, QR, RS, SB being straight and lying entirely in the liquid.

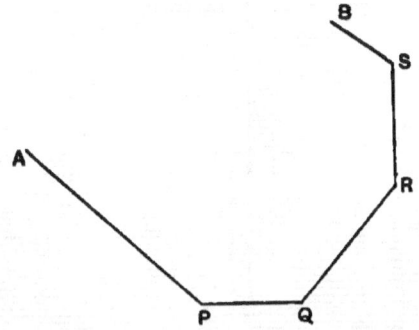

By (α) the pressure at A = that at P,
= that at Q,
= ………
= ……… B.

It should be observed that the above theorem holds for gases and for fluids of varying density.

21. Prop. *In a homogeneous fluid at rest under the action of gravity, the difference of pressure at two points varies as the difference of their depths.*

Let P, Q be the two points.

We shall consider two cases:

(α) When the straight line PQ is vertical and entirely in the liquid.

About PQ as axis construct a cylinder of indefinitely small section α.

Let $p =$ the pressure at P, p' that at Q; let w be the weight of the unit volume of the fluid.

Consider the equilibrium of the fluid contained within the cylinder.

The forces acting on it are

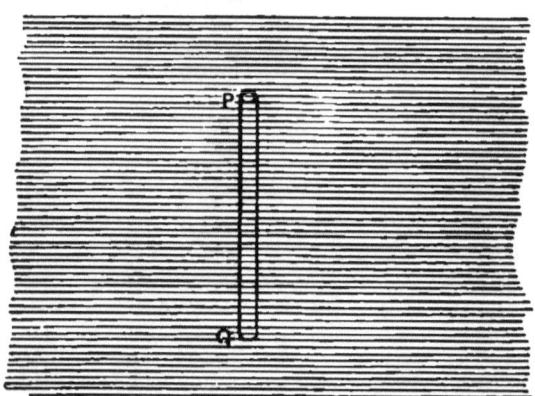

(i) the thrusts on the curved surface, everywhere horizontal,

(ii) its weight, vertically downwards, $PQ \cdot \alpha w$,

(iii) the thrust on the end at P, vertically downwards, $p\alpha$,

GENERAL THEOREMS RELATING TO PRESSURE. 25

(iv) the thrust on the end at Q, vertically upwards, $p'a$.

∴ resolving vertically

$$p'a = pa + PQ \cdot aw,$$
$$\therefore p' - p = PQ \cdot w = (h' - h) w,$$

if h, h' are the depths of P, Q below any given horizontal plane.

(β) When the straight line PQ is not vertical, and does not lie entirely in the fluid.

We can join PQ by a series of straight lines PA, AA', $A'B$, BB', $B'C$, CC', $C'Q$, each of which is entirely in the fluid, and either vertical or horizontal.

As we move along the horizontal lines AA', BB', CC', the pressure does not alter; and by case (α) as

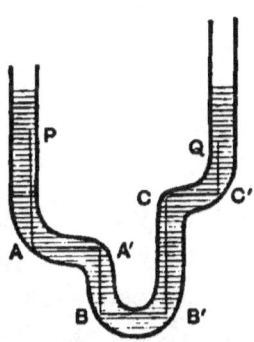

we move down PA, $A'B$, the pressure is increased by $(PA + A'B)w$, and as we move up $B'C$ and $C'Q$, it is decreased by $(B'C + C'Q)w$;

∴ the pressure at Q — that at P

$$= w (PA + A'B - B'C - C'Q)$$
$$= (h' - h) w = g\rho (h' - h),$$

if $\rho =$ the density of the liquid.

26 GENERAL THEOREMS RELATING TO PRESSURE.

Cor. Hence in a homogeneous fluid the pressure at two points at the same level is the same, even when they cannot be joined by a horizontal line, curved or otherwise, lying entirely in the fluid.

Conversely, if the pressures at two points in a homogeneous fluid be the same, they must be at the same level.

22. Prop. *The densities at two points in a fluid at rest under gravity and in the same horizontal plane are equal.*

(It is assumed that the two points can be joined by a horizontal line, curved or otherwise, lying entirely in the liquid.)

Let P, Q be the two points.

Draw PP', QQ' vertically downwards, and of the same indefinitely small length. Then P', Q' are in the same horizontal plane.

Since PP' is indefinitely small, the densities at P and P' differ by an indefinitely small quantity of the order PP'; hence the fluid between them may be regarded as homogeneous, and similarly for the fluid between Q and Q'.

Hence if ρ, ρ' be the densities at P, Q respectively,

the pressure at P' — that at $P = g\rho \cdot PP'$ (Art. 21).

Also .... Q' — .... $Q = g\rho' \cdot QQ'$.

But the pressure at $P' =$ that at Q' (Art. 20),

and .... $P = $......... Q;

$$\therefore g\rho \cdot PP' = g\rho' \cdot QQ',$$

$$\therefore \rho = \rho'.$$

GENERAL THEOREMS RELATING TO PRESSURE. 27

Cor. *The surface of separation between two fluids of different densities is a horizontal plane.*

For if the surface is not a horizontal plane a horizontal straight line PQ can be drawn cutting the surface, so that P is on one side the surface and Q on the other, i.e. the density at P is different from that at Q, which is impossible by the above proposition.

As particular cases of the above, we may take the surface of a liquid with an atmosphere above it or with merely its own vapour above it. It is of course assumed that both fluids are at rest; if one, the atmosphere for instance, were in motion, the surface of separation would not of necessity be a horizontal plane.

It follows from the above Cor., that if a homogeneous liquid at rest have a number of isolated surfaces in contact with the *same* atmosphere at rest, the surfaces must lie in the same horizontal plane. This fact is popularly expressed by the saying that 'water always seeks its own level.' This is illustrated by the experimental fact that, if water be conveyed by closed pipes to a town from a reservoir outside, it will always rise to the level of the surface in the reservoir if free to do so, but no higher.

23. Prop. *To determine the pressure at any depth in a homogeneous liquid in contact with an atmosphere at rest.*

By Art. 22, the surface of the liquid is a horizontal plane; let Π be the pressure of the atmosphere at the surface, and p that at a point P, at a depth h below the surface. Let ρ be the density of the liquid.

By Art. 21, $\quad p - \Pi = g\rho h,$
$$\therefore p = \Pi + g\rho h.$$
If there is no atmospheric pressure,
$$p = g\rho h,$$
i.e. *the pressure at any point varies as the depth below the surface when there is no atmospheric pressure.*

When Π is not zero, let us imagine the atmosphere to be removed, and a stratum of the liquid of thickness $\Pi/g\rho$ ($= h'$ say) to be placed above the original liquid. Then the pressure at a point at a depth h below the original surface is $g\rho(h + h')$, i.e. $g\rho h + \Pi$, or the same as it was before the atmosphere was removed.

The upper surface of the superimposed liquid is termed the *Effective* surface, and hence *the pressure at any point of a homogeneous liquid is proportional to the depth below the effective surface.*

It is on this account that the pressure at any point is often said to be due to such a depth of liquid, or to such a '**head**' of liquid, meaning that it is at that depth below the effective surface. Thus we may say that a certain pressure is that due to a head of 60 feet of water, meaning that it is the same as the pressure at a point in water 60 feet below the effective surface.

EXAMPLES.

1. The pressure in the water-pipe at the basement of a building is 34 lb. wt. to the square inch, whereas at the third floor it is only 18 lb. wt. to the square inch. Find the height of the third floor.

2. Water being the standard substance, find the units of length and time that the formulæ $p = g\rho z$, $p = \sigma z$, may both give the pressure at a depth z in ounce weights. [M. T., 1865.]

3. Find the pressure in lbs. wt. per sq. foot at a point of the base of a cylinder whose section is 5 sq. ft., and which contains mercury (sp. gr. 13·6) to a depth of 3 inches, water to a depth of 18 inches, and oil (sp. gr. ·65) to a depth of 6 inches, the atmospheric pressure being 15 lbs. to a sq. inch.

4. Prove that if a parallelogram be immersed in any manner in a heavy homogeneous fluid, the sum of the pressures at the extremities of one diagonal is equal to the sum of the pressures at the extremity of the other diagonal. [M. T., 1871.]

GENERAL THEOREMS RELATING TO PRESSURE. 29

5. The lower ends of two vertical tubes whose cross sections are 1 and ·1 square inches respectively are connected by a tube. The tubes contain mercury (sp. gr. 13·6). How much water must be poured into the larger tube to raise the level of the mercury in the smaller tube by one inch?

6. A right circular cone whose height is 1 foot, area of base 2 sq. ft., contains 18 cu. inches of mercury, 126 of water, and 342 of oil (v. Ex. 3); find the pressure at the lowest point, when it is placed with axis vertical, and the vertex downwards, the atmospheric pressure being 15 lbs. per sq. inch.

7. Three fluids, whose densities are in A.P., fill a semicircular tube whose bounding diameter is horizontal. Prove that the depth of one of the common surfaces is double that of the other. [M. T., 1861.]

8. If mercury be taken as the standard of density and a yard and minute as the units of space and time, find the depth below the surface of mercury at which the pressure is unity (a cubic inch of mercury weighs nearly $\frac{1}{2}$ lb.) and express the unit of pressure in lbs. weight per sq. inch.

9. Two uniform vertical tubes are connected by a fine horizontal one, and mercury is poured in until it occupies the horizontal tube, and a foot of each vertical tube. Water is now poured into one tube and glycerine into the other, so as to occupy lengths of 2 ft. and 4 ft. respectively. Find the point in the other tube where the pressure is the same as at the mercury-glycerine surface.

Specific gravity of mercury $=13\cdot6$, that of glycerine $=1\cdot25$.

10. Two indefinitely long cylinders equal in all respects are placed on a horizontal plane, their bases being connected by a horizontal pipe of small section. The cylinders are filled to depths of 2 ft. and 1 ft. by fluids of densities 3ρ and 6ρ respectively. Shew that, if a volume of fluid of density 4ρ, which would occupy a length of either cylinder equal to 3 ft. be poured into the second cylinder, the free surfaces of the fluids in the cylinders will lie in the same horizontal plane. [Jesus Coll., 1883.]

11. If ρ, ρ' be the densities of two fluids ($\rho < \rho'$), and the lengths of the arms of a U-tube in which they meet be m and n inches respectively: prove that in order that the tube may be completely filled, the height of the column of the lighter fluid above the horizontal plane in which they meet must be $\rho'(m-n)/(\rho'-\rho)$ inches. [M. T., 1859.]

12. A small uniform tube is bent into the form of a circle whose plane is vertical, equal volumes of two fluids whose densities are ρ, σ fill

half the tube; shew that the radius passing through the common surface makes with the vertical the angle $\tan^{-1}(\rho - \sigma)/(\rho + \sigma)$.

[Jesus Coll., 1886.]

13. A hollow cone, whose axis is vertical and base downwards, is filled with equal volumes of two liquids, whose densities are in the ratio of $3:1$; prove that the pressure at a point in the base is $(3 - \sqrt[3]{4})$ times as great as when the vessel is filled with the lighter fluid. [M. T., 1870.]

14. A uniform tube is bent into the form of a cycloid and held with its vertex downwards and its axis vertical. It is then partly filled with mercury (specific gravity 13·5) and chloroform (specific gravity 1·5). Shew that, if the volume of the chloroform be three times that of the mercury, their common surface will be at the lowest point of the tube.

[Jesus Coll., 1887.]

15. A closed tube in the form of an equilateral triangle contains equal volumes of three liquids which do not mix, and is placed with its lowest side horizontal. Prove that, if the densities of the liquids be in arithmetical progression, their surfaces of separation will be at points of trisection of the sides of the triangle. [M. T., 1874.]

16. In the lower half of a uniform circular tube, one quadrant is occupied by a liquid of density 2ρ, and the other quadrant is occupied by two liquids, which do not mix, of densities ρ and 3ρ; prove that the volume of the lower of the two latter liquids is twice that of the upper.

[M. T., 1868.]

17. A thin uniform cycloidal tube contains equal weights of two fluids: if it be placed with its axis vertical, prove that the heights of the free surfaces of the fluids above the vertex of the tube are as

$$(3a+b)^2 : (3b+a)^2,$$

where a and b are the lengths of the tube which the fluids occupy.

[M. T., 1867.]

*24. We shall now prove more general theorems analogous to those which we have proved in Arts. 20—22 for fluids at rest under gravity.

DEF. **Surfaces of equal pressure.** *The surface passing through all the points of a fluid at which the pressure is the same is a surface of equal pressure.*

We have seen (Art. 20) that in a fluid at rest under the action of gravity, a horizontal plane is a surface of equal pressure, and also one of equal density (Art. 22).

DEF. *A line of force is a line whose direction at every point coincides with that of the resultant external force.*

When gravity is the only external force, the lines of force are vertical.

*25. PROP. *In a fluid at rest the surfaces of equal pressure cut the lines of force at right angles.*

Let A be any point in the fluid. From A draw an indefinitely short line AB in the surface of equipressure through A. About AB as axis construct a

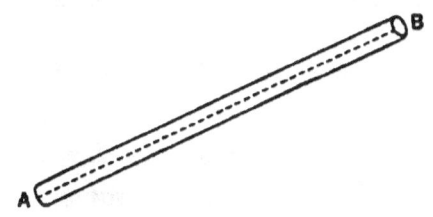

circular cylinder, whose radius is indefinitely small compared with AB.

(The radius is taken indefinitely small compared with AB in order that when we take the pressure at P to be the mean pressure over the end of the cylinder there, the error, which depends on the radius of the end, may be small compared with AB.)

Since the pressure at $A =$ that at B, the thrust on the end at $A =$ that on the end at B.

Therefore the resultant thrust on the cylinder is perpendicular to AB. But (Art. 13) the resultant thrust is along the line of force. Therefore the line of force is perpendicular to AB.

Similarly we can shew that the line of force at A is perpendicular to any line through A in the surface of equi-pressure there.

Hence the surface of equal pressure at A is perpendicular to the line of force there.

The proposition of Art. 20 is clearly a particular case of the above theorem, in which the surfaces of equal pressure are horizontal planes at right angles to the vertical lines of force.

EXAMPLES.

1. Prove that in a fluid at rest under the action of a force towards a fixed point, the surfaces of equal pressure are concentric spheres.

2. A uniform liquid rests in equilibrium in a vessel under the action of gravity and of a force towards a fixed point at the bottom of the vessel varying as the distance from that point: shew that the free surface is a portion of a sphere. [M. T., 1886.]

*26. Prop. *To shew that the distance between consecutive surfaces of equal pressure varies inversely as the resultant force and the density, conjointly.*

Let P be a point on a surface of equal pressure (p).

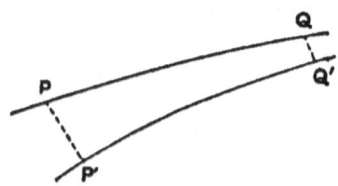

Draw the normal PP' to meet the consecutive surface of equal pressure (p') at P'.

About PP' as axis construct a circular cylinder of radius indefinitely small compared with PP'.

Consider the equilibrium of the fluid contained within

GENERAL THEOREMS RELATING TO PRESSURE. 33

this cylinder. Let ρ be the density, f the external force per unit mass in direction PP', and α the area of the section of the cylinder.

The forces acting on the cylinder are

(i) $f\rho\alpha \cdot PP'$, the external force along PP',
(ii) $p\alpha$, the thrust on the end at P along PP',
(iii) $p'\alpha$, the thrust on the end at P' along $P'P$,
(iv) the thrusts on the curved surface, at right angles to PP'.

Hence resolving along PP',
$$p\alpha + f\rho\alpha \cdot PP' - p'\alpha = 0;$$
$$\therefore p' - p = f\rho \cdot PP'.$$

If ds be the length of the small arc PP' measured along the line of force this result may be expressed thus,
$$\frac{dp}{ds} = f\rho.$$

The proposition of Art. 21 is a particular case of this.

We can prove in a similar manner that $p' - p = f\rho \cdot PP'$, where PP' is the distance measured in *any* given direction between the consecutive surfaces of equal pressure p, p', and f is the resolved part of the external force in the same direction.

Thus if X be the resolved part of the external force along any axis Ox, and dx be the length PP', this result may be written
$$\frac{dp}{dx} = \rho X.$$

EXAMPLES.

1. Prove that, if the density of a liquid at rest under gravity varies as the square root of the pressure, the density increases uniformly with the depth. [M. T., 1884.]

2. If it be assumed that the earth is a sphere, and that the attraction it exerts on an internal point varies as the distance from the centre, prove that the pressure at any point in the sea varies as the rectangle contained by the segments of a line through the point. [Jesus Coll., 1889.]

3. If a liquid be heterogeneous and of density $\rho z/a$ at a depth z, shew that the pressure is $\Pi + g\rho z^2/2a$, where Π is the atmospheric pressure. [Pet., 1885.]

4. Shew that the pressure at a small depth z below the surface of a sphere of water attracted to the centre of the sphere with a force producing an acceleration μ/r^2 at a distance r is approximately
$$\Pi + g\rho\,(z + z^2/a),$$
where a is the radius of the sphere and g the attraction on a unit mass at the surface of the sphere. [Pet., 1887.]

*27. Prop. *When a fluid is in equilibrium under the action of a conservative system of forces, the surfaces of equal density coincide with those of equal pressure.*

Let P, Q, as in the last Article, be any two points on a surface of equi-pressure (p).

Let f, ρ be the resultant force and density at P, f', ρ' the same at Q.

Draw the normals PP', QQ' to meet the consecutive surface (p').

Since the external forces form a conservative system, the work done on a particle of unit mass as it describes the circuit $PQQ'P'$ is zero.

The work done in moving along PQ and $Q'P'$ is zero, since the path is everywhere at right angles to the lines of force.

The work done in moving along $QQ' = f' \cdot QQ'$,

$$P'P = -f \cdot PP';$$

$$\therefore f' \cdot QQ' - f \cdot PP' = 0.$$

But by the last Article,

$$f\rho \cdot PP' - f'\rho' \cdot QQ' = 0,$$

$$\therefore \rho = \rho';$$

∴ the density is uniform over the surface of equi-pressure.

If the fluid be such that the density depends solely on the temperature and the pressure, the temperature also must be uniform over a surface of equi-pressure, i.e. the surfaces of equi-pressure, -density, and -temperature coincide.

The theorem of Art. 22 is clearly a particular case of this one.

COR. *The surface of separation between two fluids of different densities, at rest under the action of a conservative system of forces, is a surface of equi-pressure.*

The proof is similar to that of the Corollary to Art. 22.

*28. The converse proposition to the last also holds, viz. *if the surfaces of equal pressure are also surfaces of equal density, in a fluid in equilibrium, the external forces must form a conservative system.*

Consider any closed circuit, cutting the surface of equi-pressure (p) in P and Q, and the consecutive one (p') in Q' and P'.

The work done on a particle of unit mass in describing the elements $P'P$, QQ' of the path are $-f \cdot P'P$, $f' \cdot QQ'$ respectively, where f, f' are external forces at P, Q in directions PP', QQ' respectively.

But (Art. 26) $f\rho . PP' = p' - p = f'\rho' . QQ'$, and by hypothesis $\rho = \rho'$,

$$\therefore -f . PP' + f' . QQ' = 0.$$

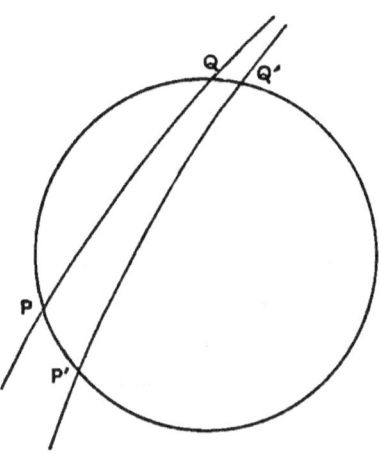

The whole circuit can be divided up into pairs of elements similar to PP' and QQ', such that the work done in describing any pair is zero: therefore the total work done in describing the whole circuit is zero; the system of external forces must therefore be a conservative one.

The following is a particular case of this theorem, *a homogeneous liquid cannot be in equilibrium under the action of a non-conservative system of forces.*

*29. PROP. *A homogeneous liquid completely filling a closed vessel and under the action of a conservative system of forces must be in equilibrium.*

For, if not, the liquid will begin to move so that the work done on it is positive, i.e. so that its Potential Energy is diminished. But since, considered as a whole, the liquid occupies exactly the same position as at first, its Potential Energy is unaltered, i.e. it must be in equilibrium.

GENERAL THEOREMS RELATING TO PRESSURE.

*30. The propositions of Arts. 25—29 have been deduced from the characteristic property of all fluids at rest, whether viscous or otherwise, that the stress across any surface is everywhere normal. In the case of perfect fluids and, for certain kinds of motion, of viscous fluids also (Art. 12), we can prove analogous theorems for fluids in motion. When an element of fluid is at rest, we have seen (Art. 13) that it is acted on by two equal and opposite forces, the resultant external force and the resultant thrust: when it is in motion, the resultant of these two forces gives the element its resultant acceleration.

Thus if at a point P of the liquid, where the pressure is p, the density ρ, the external force along a fixed line Ox, X, and the acceleration along Ox, α, we can shew that instead of the result of Art. 26, we have

$$\frac{dp}{dx} = \rho(X - \alpha).$$

*31. Prop. *The direction of the resultant thrust on an element of a perfect fluid in motion, all of whose dimensions are indefinitely small, is perpendicular to the surface of equal pressure through the element.*

The direction of the resultant thrust is independent of the *shape* of the element, since it depends on the resultant force and the resultant acceleration only, and both of these are independent of the shape of the element.

Take an indefinitely small cylindrical element, whose axis PQ lies in a surface of equal pressure. The thrusts on the ends being equal and opposite, the resultant thrust must be perpendicular to PQ. In a similar way it can be shewn that the resultant thrust is perpendicular to any line in the surface of equal pressure. The required result therefore follows.

EXAMPLES.

1. Two equal and in every respect similar buckets (weight W) are connected by a cord which passes over a smooth pulley. If a weight, $3W$, of water is poured into one bucket and a weight W into the other, compare the pressure of the water at the base of either bucket with what it would be if the bucket were at rest.

2. A closed vessel containing water and mercury is moved downwards with acceleration f: determine the surface of separation between the two liquids, and the pressure at any point when f is (1) $<g$, (2) $>g$.

3. If a vessel in which water is contained slide down a smooth inclined plane, find the inclination to the horizon of the surface of the fluid when at rest relative to the vessel. Find the pressure at a given distance from the surface.

32. We shall apply the last proposition to determine the surfaces of equi-pressure in a fluid revolving with uniform angular velocity about a vertical axis. In this case, if the vessel holding the fluid also revolve about the vertical axis with the same angular velocity, there is no tendency for one portion to slide over another, and the proof will apply to a viscous fluid.

PROP. *To determine the surfaces of equi-pressure in a heavy fluid revolving with uniform angular velocity about a vertical axis.*

Let GN be the vertical axis, ω the angular velocity.

Let m be the mass of an element of the fluid, situate at P. This element is acted on by its weight mg vertically downwards and the resultant thrust, which together give it the acceleration $\omega^2 PN$ along PN. Their resultant must therefore by Newton's Second Law be $m\omega^2 PN$ along PN.

GENERAL THEOREMS RELATING TO PRESSURE. 39

Mark off NG along the axis above N, so that $NG = \dfrac{g}{\omega^2}$, and join PG.

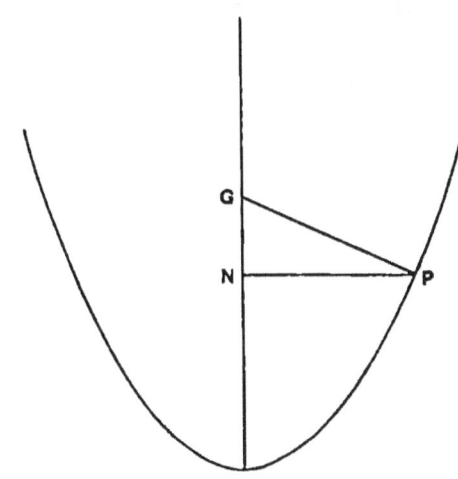

$$\therefore\ GN : NP = g : \omega^2 PN,$$

the sides GN, PN of the triangle GNP represent the forces mg, and $m\omega^2 PN$. Hence by the triangle of forces PG must represent the resultant thrust. Therefore (Art. 31) PG is the normal to the surface of equal pressure through P, and NG the subnormal is constant, (g/ω^2).

As the normal always intersects the axis NG, the surface of equi-pressure must be one of revolution about NG, and as the subnormal is constant and above N, the surface must be generated by the revolution of a parabola whose latus rectum is $2g/\omega^2$, and whose axis is along NG, with the vertex downwards. All the surfaces of equi-pressure then are equal paraboloids of revolution about the axis of rotation, and with vertices one below another.

33. Prop. *The surfaces of equi-pressure in the last Article are also surfaces of equal density.*

Let P, Q be two points on a surface of equi-pressure (p). Let ρ, σ be the densities at P, Q respectively. Draw PP', QQ' vertically downwards to meet the consecutive surface of equi-pressure (p') in P', Q', respectively.

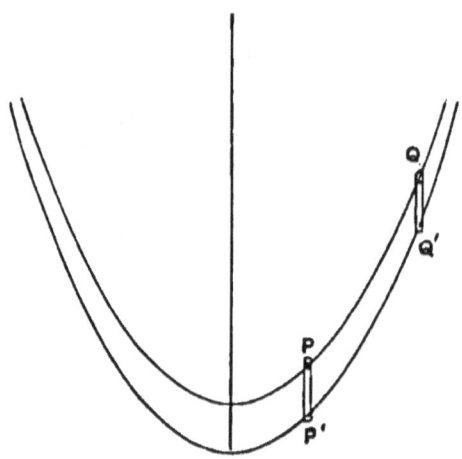

Construct a cylinder of radius indefinitely small compared with PP' about PP' as axis, and consider the fluid contained within it.

The only vertical forces are the thrusts on the ends at P and P' and the weight, and there is no vertical motion. Hence if a is the sectional area of the cylinder,
$$p'a = pa + g\rho a \cdot PP',$$
$$\therefore p' - p = g\rho \cdot PP'.$$
Similarly $\qquad p' - p = g\sigma \cdot QQ'.$
But $\qquad PP' = QQ';$
$$\therefore \rho = \sigma,$$

GENERAL THEOREMS RELATING TO PRESSURE. 41

i.e. the surfaces of equi-pressure are also surfaces of equal density.

COR. *If there are two homogeneous fluids of different density which do not mix, their surface of separation is one of the family of paraboloids.*

For it is clear that this surface cannot *cut* a surface of equal density, and must therefore itself be one of the paraboloids.

A particular case of this is the free surface of a liquid rotating with uniform angular velocity. It will be the highest paraboloid of the series.

34. *To find the pressure at any point of a homogeneous liquid revolving with uniform angular velocity about a vertical axis.*

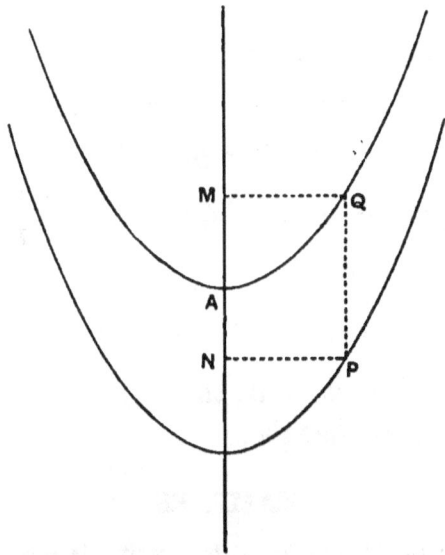

Let Π be the external pressure which we assume to be uniform. Let AN be the vertical axis, ω the angular velocity about it, ρ the density of the liquid.

By Art. 33, the free surface is the highest of the series of paraboloids, which form the surfaces of equi-pressure.

Let A be the vertex of this paraboloid. Let P be any point in the liquid.

Draw PQ vertically upwards to meet the free surface in Q; draw PN, QM perpendicular to AN.

About PQ as axis construct a cylinder of indefinitely small sectional area α, and consider the liquid contained in it.

As there is no vertical motion, the vertical forces must balance one another.

These vertical forces are

(i) thrust on end at Q, $\Pi \alpha$,

(ii) .... P, $p\alpha$,

(iii) the weight $g\rho\alpha . PQ$,

$$\therefore p\alpha = \Pi\alpha + g\rho\alpha . PQ,$$
$$\therefore p = \Pi + g\rho . PQ.$$

But $PQ = AN + AM = AN + QM^2 . \dfrac{\omega^2}{2g}$

$$= AN + PN^2 . \omega^2/2g;$$
$$\therefore p = \Pi + g\rho AN + \tfrac{1}{2} PN^2 . \omega^2 \rho.$$

NOTE. If N is *above* A, the sign of AN in the above expression must be changed.

EXAMPLES.

1. A uniform liquid, unaffected by gravity, is revolving in relative equilibrium with uniform angular velocity about an axis. Find the form of the surfaces of equal pressure, and the pressure at any point.

2. A vessel in the form of a right cone with its vertex downwards is filled with liquid and revolves with uniform angular velocity round the

axis: having given the height of the cone and the vertical angle, determine how much liquid will remain in relative equilibrium for an assigned angular velocity. [M.T., 1866.]

3. If the cylinder and its contents in Ex. 3, p. 28, revolve with uniform angular velocity ω about its axis, supposed vertical, find the pressure at any point of its base.

4. Fluid is rotating in a cylinder of radius r, whose bottom is closed by a conical surface of semi-vertical angle a, the vertex being downwards. Prove that the pressure at the surface of the cone is a minimum at a point distant $\frac{1}{2}l \cot a$ from the axis where l is the latus rectum of the free surface, provided $l < 2r \tan a$. [Pet., 1886.]

THRUST ON A PLANE AREA.

35. Prop. *The thrust on any plane area exposed to a homogeneous liquid under gravity is equal to the weight of a column of the liquid whose base is equal to the area, and whose height is equal to the depth of the centre of mass of the area below the effective surface of the liquid.*

Let the area A be divided into an infinite number of indefinitely small areas $a_1, a_2, a_3, \ldots$, and let the depths of these areas below the free surface be $z_1, z_2, z_3, \ldots$: let Π be the external pressure and ρ the density of the fluid.

$\bar{z}$, the depth of the C.M. of the surface,

$$= \frac{z_1 a_1 + z_2 a_2 + z_3 a_3 + \ldots}{A}.$$

The pressure at any point of $a_1 = \Pi + g\rho z_1$,

$\ldots\ldots\ldots\ldots\ldots\ldots\ldots\ldots\ldots\ldots\ldots\ldots\ a_2 = \Pi + g\rho z_2,$

$= \ldots\ldots$

$\therefore$ the thrust on A

$= (\Pi + g\rho z_1) a_1 + (\Pi + g\rho z_2) a_2 + \ldots$

$= \Pi (a_1 + a_2 + \ldots) + g\rho (z_1 a_1 + z_2 a_2 + \ldots)$

$= (\Pi + g\rho \bar{z}) A = g\rho z' A,$

where z' is the depth of the C.M. below the effective surface.

COR. *The mean pressure throughout the area is equal to the pressure at its centre of mass.*

36. It has been usual in text-books on Hydrostatics to define the '*whole pressure*' on any surface as the numerical sum of the thrusts on the infinite number of indefinitely small plane areas into which the surface may be divided; and to prove that this whole pressure is equal to the area of the surface multiplied by the pressure at its centre of mass. This may be proved by the method of the last Article. It has been purposely omitted here as the whole pressure on a *curved* surface has no physical meaning.

EXAMPLES.

1. A regular hexagon is placed with its plane vertical and its centre at a depth c in a liquid of density ρ. If $a\,(<c)$ be the side of the hexagon, and Π be the external pressure, find the thrusts on the triangles into which the hexagon is divided by joining the corners with the centre.

2. A rectangle whose sides are 10 ft. and 15 ft. respectively is placed with its plane vertical and one of its shorter sides in the surface of some water, which is 14 ft. deep, and which rests on a stratum of mercury (sp. gr. 13·6). Find what is the volume of water whose weight is equal to the thrust on the rectangle, if the atmospheric pressure is equivalent to a head, 33 ft. of water.

3. The lighter of two fluids, whose specific gravities are as 2 : 3, rests on the heavier, to a depth of 4 inches. A square is immersed in a vertical position with one side in the upper surface: determine the side of the square in order that the thrusts on the portions in the two fluids may be equal. [M. T., 1855.]

4. A triangle is immersed in a homogeneous liquid with one side in the surface: shew how to draw a horizontal line dividing it into two portions on which the thrusts are equal.

5. A parallelogram is immersed in a homogeneous fluid with one side in the surface; shew how to draw a line from one extremity of this side dividing the parallelogram into two parts the thrusts on which are equal. [M. T., 1860.]

GENERAL THEOREMS RELATING TO PRESSURE. 45

6. The side AB of a triangle ABC is in the surface of a fluid, and points D, E are taken in AC, such that the pressures on the triangles BAD, BDE, BEC are equal: find the ratios $AD : DE : EC$.

[M. T., 1856.]

7. A hollow weightless hemisphere, filled with liquid, is suspended freely from a point in the rim of its base; find the thrust on the base.

[St John's, 1887.]

8. A parallelogram is immersed in a homogeneous liquid with one side in the surface: shew how to draw horizontal lines dividing it into n portions the thrusts on which are equal.

9. Shew that the thrust on the base of a vessel containing a homogeneous liquid depends solely on the depth of the liquid and the area of the base, being independent of the shape of the vessel.

IMPULSIVE PRESSURE.

*37. If the motion of a mass of liquid be suddenly changed, the internal forces thereby set in action will be what are ordinarily termed *impulsive*, and are measured by the *change* of momentum produced, and not by the *rate of change*. Thus *the impulsive pressure at any point is the limit of* I/A, *where* A *is an indefinitely small area containing the point and* I *is the normal impulse across it.*

It is easily seen that impulsive pressure is of 1 dimension in mass, -1 in length, and -1 in time.

*38. Since in a perfect fluid there is never any shearing stress, the impulsive pressure at any point can be shewn, as in Art. 11, to be the same in all directions.

In all cases where motion is suddenly changed, forces such as gravity, which require an appreciable time to produce an appreciable effect, are neglected, so that the only forces acting on any mass of fluid at the moment of the impulse, are the impulsive thrusts. By considering the change of momentum produced in elements of fluid of the corresponding shapes, the following propositions analogous to those of Arts. 25 and 26 respectively can be proved.

(a) *If the motion of a perfect fluid be suddenly changed, the surface of equal impulsive pressure at any point is perpendicular to the direction of the change of motion there.*

(b) *If* v *be the change of velocity produced at any point of a perfect fluid in a given direction* Ox, *and* ρ *be the density there*

$$\frac{dp}{dx} = \rho v$$

where p *is the impulsive pressure at the point.*

As a particular case of the latter proposition we may take the case of a vessel containing liquid, which is moving vertically downwards with velocity v and is suddenly stopped. The impulsive pressure at a depth x below the surface is $\rho v x$. Hence, as in Art. 35, it can be shewn that the resultant impulsive thrust on any plane area A is $\rho v \bar{x} A$, where $\bar{x}$ is the depth of A's centre of mass below the surface.

It can also be shewn, as in Art. 27, that if a perfect liquid have a uniform velocity in any direction and be suddenly stopped, the surfaces of equal density, as well as those of equal impulsive pressure, are perpendicular to the direction of the velocity. As these conditions are not always satisfied, it follows that *any* given motion of a liquid cannot always be suddenly annihilated.

Ex. Shew that, if a vessel partly full of liquid and moving uniformly along a horizontal plane be suddenly stopped, the liquid is not at once reduced to rest.

GENERAL THEOREMS RELATING TO PRESSURE.   47

39. In Art. 22, it was deduced from the definition of a fluid given in Art. 4, that the surface of a liquid at rest is a horizontal plane. It may be instructive to shew the converse, viz. that if we assume as an observed fact that the surface of a liquid at rest is always horizontal, it must follow that in a homogeneous liquid at rest no tangential stress is exerted.

Imagine a hollow cylinder, open at one end, and of indefinitely small section to be placed anywhere in a homogeneous liquid at rest. Let a close-fitting frictionless piston work inside this cylinder, and be kept in equilibrium by the thrust of the liquid on one end and by that of a spring inside the cylinder on the other. Now suppose that the spring becomes slightly weaker so that the piston gives way slowly to a small extent under the thrust of the liquid. We know from observation that the liquid will flow in to fill up the gap left by the piston and that the surface of the liquid will in consequence be lowered. The total work done by the forces, external and internal, on the liquid during this displacement is zero. The external work done is (1) that by gravity in lowering the liquid from the surface to the gap, (2) that done by the external pressure on the surface of the liquid, and (3) that done *against* the thrust of the spring. The internal work is that done against the shearing forces which resist one portion sliding over another. Now we may allow the liquid to flow from one part to another by the shortest course, or, by inserting a number of fixed diaphragms, we may compel the liquid to take a course as intricate as we please. Whichever is done, the external work remains the same, and therefore the internal work also. But if there are any shearing stresses, the work done against them must be greater when the course the liquid takes is a very long one, than when it is the shortest possible. We conclude then that there are no tangential stresses.

In the above case, let a be the sectional area of the piston, h the distance it is moved, z its depth below the surface, and p the pressure at the end of it. Let ρ be the density of the liquid, Π the external pressure, and A the area of the free surface.

If x is the distance the free surface is lowered when the spring gives way, $xA = ha$, since a volume equal to that of the gap is transferred from the surface to fill up the gap.

The work done by gravity $= g\rho haz$,

.... external pressure $= A\Pi x = \Pi ha$,

.... against resistance of spring $= pah$.

$\therefore g\rho haz + \Pi ha = pah$,

$\therefore p = \Pi + g\rho z$,

the result obtained in Art. 23.

48 GENERAL THEOREMS RELATING TO PRESSURE.

Also, since nothing has been said about the inclination of the end of the piston to the horizon, the result is independent of it, and hence the pressure is the same in all directions.

We can also deduce in a similar way the result of Art. 35.

For let A be the area: let another area be drawn parallel to it so that the distance between the two is everywhere h, an indefinitely small quantity. As before, let one area slowly give way before the liquid pressure on it until it coincides with the other, the liquid flowing in from the free surface to fill up the gap. As before the total work done on the liquid is zero.

The work done by gravity in lowering the weight $Ahg\rho$ from the free surface to fill up the gap is $Ahg\rho \cdot \bar{z}$, where $\bar{z}$ is the depth of the centre of mass of A. That done by the external pressure is $\Pi \cdot Ah$, and that done against the resistance of the area is Ph, where P is the thrust.

$$\therefore Ahg\rho\bar{z} + \Pi Ah = Ph;$$
$$\therefore P, \text{ the thrust on } A = (g\rho\bar{z} + \Pi) A.$$

ILLUSTRATIVE EXAMPLES.

1. *A fine tube ABC, of uniform bore, having the parts AB, BC straight and perpendicular to one another, is held in a vertical plane and contains several liquids of different densities. If the tube be turned about the point B in its own plane and the liquids be again allowed to settle without spilling, prove that the weight of liquid which has passed from one branch to the other bears to the weight of the whole the ratio $\tan a \sim \tan \beta : 2$, where a, β are the inclinations to the vertical of the straight line bisecting the angle ABC in the two positions.* [M. T., 1870.]

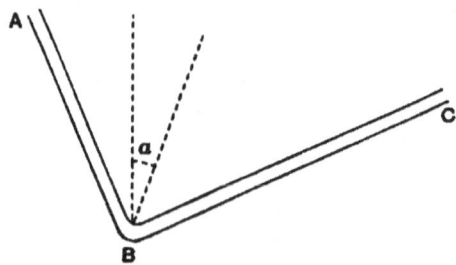

Let W_1, W_2 be the weights of liquid in AB, BC respectively in the first position; and let W be the amount that flows from AB to BC when the tube is turned into the second position. Let a be the section of the tube, Π the external pressure on the two surfaces. Let p be the pressure at B in the first position.

Considering the equilibrium of the column of liquid in AB, and resolving along the tube, we have

$$\Pi a + W_1 \cos(\tfrac{1}{4}\pi - a) = pa.$$

Similarly from the equilibrium of the liquid in BC,

$$\Pi a + W_2 \sin(\tfrac{1}{4}\pi - a) = pa.$$

$$\therefore W_1/W_2 = \tan(\tfrac{1}{4}\pi - a) \quad \dots \dots \dots \dots \dots \dots \text{(i)}.$$

Similarly, when the tube is turned into the second position,

$$(W_1 - W)/(W_2 + W) = \tan(\tfrac{1}{4}\pi - \beta) \quad \dots \dots \dots \dots \text{(ii)}.$$

From (i) $\tan a = (W_2 - W_1)/(W_2 + W_1).$

From (ii) $\tan \beta = (W_2 - W_1 + 2W)/(W_2 + W_1).$

$$\therefore \tan a \sim \tan \beta = 2W/(W_1 + W_2).$$

$$\therefore W : (W_1 + W_2) = \tan a \sim \tan \beta : 2.$$

2. *If the attraction of the earth at a depth z below the surface were $a + bz$, prove that the pressure at that depth in water would be $\rho(az + \tfrac{1}{2}bz^2)$, where ρ is the density of the water.*   [Pet. 1891.]

It is assumed in the above that there is no external pressure on the surface of the water.

Let $z_1, z_2, z_3, \ldots z$ be the depths of a number of points in a vertical line, the distance between consecutive points being very small.

Let $p_1, p_2, p_3, \ldots p$ be the pressures at these points.

Then by Art. 26, $p_2 - p_1 = \rho(a + bz_1)(z_2 - z_1),$

$$p_3 - p_2 = \rho(a + bz_2)(z_3 - z_2),$$

$$\dots\dots\dots = \dots\dots\dots\dots\dots$$

$$p_{r+1} - p_r = \rho(a + bz_r)(z_{r+1} - z_r);$$

or, $p_2 - p_1 = \rho\{a + \tfrac{1}{2}b(z_2 + z_1)\}(z_2 - z_1)$

$$= \rho\{a(z_2 - z_1) + \tfrac{1}{2}b(z_2^2 - z_1^2)\},$$

since $\tfrac{1}{2}(z_2 + z_1) = z_1$, ultimately.

Similarly $p_3 - p_2 = \rho\{a(z_3 - z_2) + \tfrac{1}{2}b(z_3^2 - z_2^2)\},$

and $p_{r+1} - p_r = \rho\{a(z_{r+1} - z_r) + \tfrac{1}{2}b(z_{r+1}^2 - z_r^2)\},$

$\therefore$ adding these equations

$$p_{r+1} - p_1 = \rho\{a(z_{r+1} - z_1) + \tfrac{1}{2}b(z^2_{r+1} - z_1^2)\};$$

$$\therefore p = \rho(az + \tfrac{1}{2}bz^2),$$

$\because p_1$ and z_1 ultimately vanish.

Or, using the notation of the differential calculus,
$$dp = \rho(a+bz)\,dz.$$
$$\therefore p = \rho(az + \tfrac{1}{2}bz^2),$$
the constant introduced by integrating being zero, as p and z vanish at the surface.

3. *An open cubical box with a vertical face ABCD, containing as much water as it can hold without spilling, is moving up an inclined plane in the direction BC with uniform acceleration f. Prove that the thrust of the water on the face which passes through AB will be independent of f.*

Let a be a side of the cube, ρ the density of water, and α the inclination of the plane to the horizon.

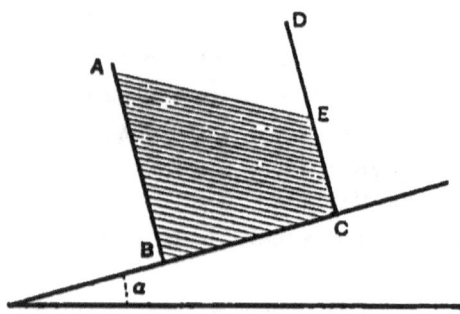

Since the box is as full as possible, the free surface passes through A.

Since there is no motion at right angles to the plane, by resolving in this direction we obtain, as in Art. 26, that the pressure at any point is $g\rho z \cos \alpha$, where z is the distance of the point from the free surface, measured perpendicular to BC.

By reasoning as in Art. 35, we see that the thrust on AB is the area of $AB \times$ the pressure at its centre of mass.

$\therefore$ the thrust on $AB = a^2 \times g\rho \cdot \tfrac{1}{2}a \cos \alpha = \tfrac{1}{2} a^3 g\rho \cos \alpha$,

which is independent of f.

4. *If a heterogeneous fluid be in stable equilibrium under the action of a conservative system of forces, the density and pressure increase together.*

Let p, ρ be the pressure and density throughout the surface of equipressure PQ: let p', ρ' be the pressure and density throughout the consecutive surface $P'Q'$. Draw PP', QQ' normals to the surface PQ. Let f, f' be the forces per unit mass along PP', QQ' respectively.

GENERAL THEOREMS RELATING TO PRESSURE. 51

Let the following indefinitely small virtual displacement be made: let a small volume v of density ρ flow from P to P' and to take its place

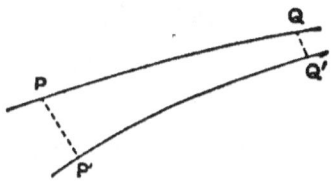

let an equal volume of density ρ' flow from Q' to Q. For stable equilibrium the work done must be negative. The work done by the internal forces is zero, and that done by the external forces

$$= f\rho v \cdot PP' - f'\rho' v \cdot QQ'$$
$$= fv \cdot PP' \cdot (\rho - \rho'),$$
$$\because f \cdot PP' = f' \cdot QQ',$$
$\therefore \rho'$ must be greater than ρ.

But since $\qquad p' - p = f\rho \cdot PP',$

p' is greater than p, i.e. p and ρ increase together.

As a particular case we see that in a heterogeneous fluid in stable equilibrium under gravity, the density must increase with the depth. If however two liquids are prevented from mixing by a flexible membrane, the heavier may be uppermost.

5. *A liquid occupies a portion of a fine circular tube of radius a, subtending an angle $\pi + \theta$ at the centre. When the tube rotates uniformly about a vertical tangent, the liquid just reaches the top of the vertical diameter: prove that the angular velocity is*

$$(g/a)^{\frac{1}{2}} (\tan \tfrac{1}{2}\theta - \sin^2 \tfrac{1}{2}\theta)^{-\frac{1}{2}}. \qquad \text{[Clare Coll., 1890.]}$$

Let NM be the vertical tangent about which the tube rotates, AOB the vertical diameter. Let ω be the angular velocity, when the liquid just rises to A. The other surface of the liquid will be at P, where the $\angle POB = \theta$.

Since A and P both lie on the free surface, they lie on a parabola, of latus rectum $2g/\omega^2$, whose axis is MN: let APK be this parabola. K being the vertex.

Draw AM, PN perpendicular to MN.

Then
$$AM^2 = \frac{2g}{\omega^2} \cdot KM,$$

and
$$PN^2 = \frac{2g}{\omega^2} \cdot KN;$$

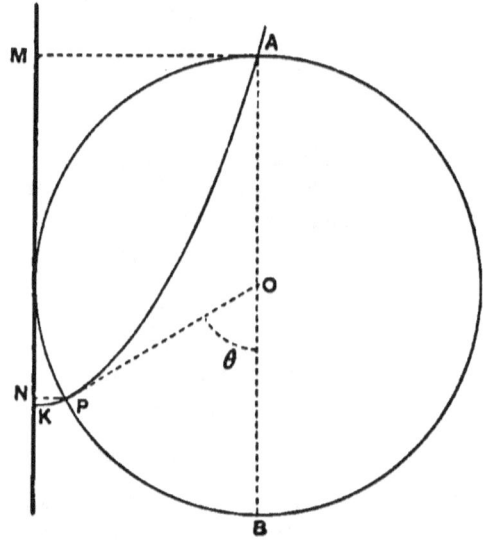

$$\therefore AM^2 - PN^2 = \frac{2g}{\omega^2} \cdot MN,$$

$$\therefore a^2 - a^2(1 - \sin\theta)^2 = \frac{2ga}{\omega^2}(1 + \cos\theta),$$

$$\therefore \omega^2 = \frac{2g}{a} \cdot \frac{1 + \cos\theta}{2\sin\theta - \sin^2\theta}$$

$$= \frac{4g\cos^2\tfrac{1}{2}\theta}{a(4\sin\tfrac{1}{2}\theta\cos\tfrac{1}{2}\theta - 4\sin^2\tfrac{1}{2}\theta\cos^2\tfrac{1}{2}\theta)},$$

$$\therefore \omega = (g/a)^{\tfrac{1}{2}}(\tan\tfrac{1}{2}\theta - \sin^2\tfrac{1}{2}\theta)^{-\tfrac{1}{2}}.$$

6. *A straight tube making an angle a with the vertical and filled with fluid of density ρ, rotates with uniform angular velocity ω about a vertical axis through its lower end which is closed. Prove that when the atmospheric pressure is Π, the greatest length of the tube that no fluid may flow out is*

$$\frac{g\rho\cos a + \omega\sin a\sqrt{(2\Pi\rho)}}{\omega^2\rho\sin^2 a}.$$
[Pet. 1886.]

GENERAL THEOREMS RELATING TO PRESSURE. 53

Let ABC be the tube, AKS the vertical line about which the rotation takes place. Through C the highest point of the liquid, draw the

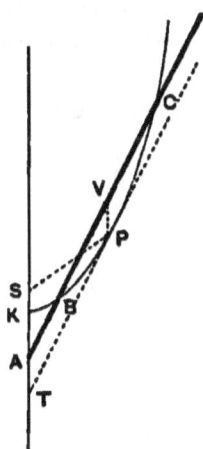

parabola KBC, having AS for axis, K for vertex, and of latus rectum $2g/\omega^2$. The pressure at every point of the liquid on this parabola is Π.

Draw PT the tangent to the parabola which is parallel to ABC. Through P the point of contact draw PV vertical, and bisecting BC in V. As the point V is at a height PV above P, the pressure there is $\Pi - g\rho \cdot PV$. The pressure at every other point of the tube must exceed this value.

Since the pressure cannot be negative, the limit to the length of the tube in order that no liquid may flow out is reached *when the pressure at V is zero*, i.e. when
$$PV = \Pi/g\rho.$$

When this condition is satisfied, the greatest length of the tube is ABC, i.e. $PT + VC$.

Let S be the focus of the parabola.

Then $\quad PT = 2SP \cos a$

$\quad\quad$ = by property of the parabola, $\dfrac{2SK \cos a}{\sin^2 a}$.

Also $\quad\quad CV^2 = 4SP \cdot PV = \dfrac{4SK}{\sin^2 a} \cdot \dfrac{\Pi}{g\rho}$.

But $\quad\quad\quad\quad SK = g/2\omega^2$.

$\therefore ABC = \dfrac{g \cos a}{\omega^2 \sin^2 a} + \dfrac{\sqrt{2\Pi\rho}}{\rho\omega \sin a}$.

7. *A closed right circular cylinder, whose axis is vertical, is very nearly filled with a homogeneous incompressible fluid. With what angular velocity must it revolve about the axis of the cylinder in order that the whole pressure on the base may be half as much again as before?*

[Jesus Coll., 1887.]

Let $ABCD$ be the cylinder, EF its axis.

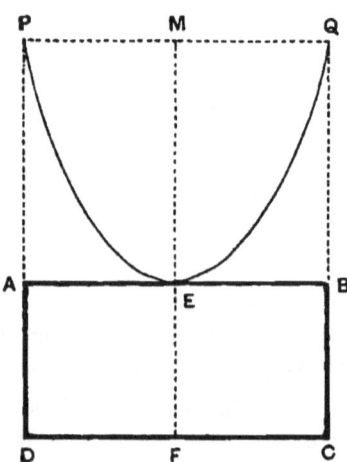

When the liquid is rotating about EF, the surface of zero-pressure will pass through E. Let PEQ be this surface. Then PEQ is a paraboloid generated by the revolution about EF of a parabola of latus rectum $2g/\omega^2$, where ω is the angular velocity.

The pressure on the base CD, when there is no rotation, is the weight of the liquid in $ABCD$; when there is rotation, the pressure is the weight of the liquid that would fill the volume $PDCQE$.

Hence if the latter pressure is half as much again as the former, the volume $PABQE$ must be half that of $ABCD$.

But vol. $PABQE$ is half vol. $PABQ$;

$$\therefore \text{vol. } PABQ = \text{vol. } ABCD,$$
$$\therefore EM = EF.$$

But
$$PM^2 = EM \cdot 2g/\omega^2,$$
$$\therefore AE^2 = 2g \cdot EF/\omega^2,$$
$$\therefore \omega^2 = 2g \cdot EF/AE^2.$$

GENERAL THEOREMS RELATING TO PRESSURE. 55

8. *To determine the surfaces of equi-pressure in a homogeneous liquid rotating as if rigid, with uniform angular velocity ω, about an axis inclined to the vertical at an angle a.*

Let AM be the axis about which the rotation takes place.

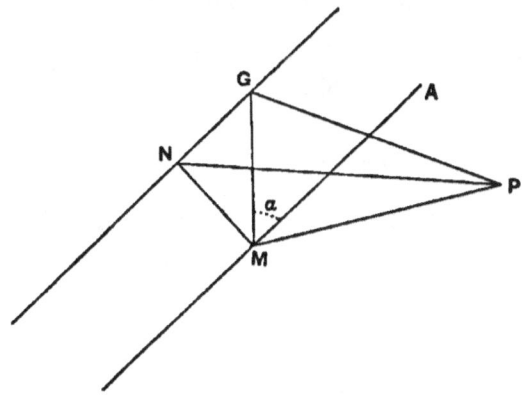

Let P be any point in the liquid: consider the motion of an element of liquid of mass m at P. Draw PM perpendicular to AM.

The forces acting on m are mg vertically downwards, and the resultant thrust along the normal to the surface of equal pressure. These give the element the acceleration $\omega^2 PM$ along PM.

Therefore their resultant is $m\omega^2 PM$ along PM.

Through M draw MG vertically upwards, so that

$$MG : MP = g : \omega^2 MP, \text{ and join } PG.$$

Then by the triangle of forces PG represents the resultant thrust, i.e. PG is normal at P to the surface of equi-pressure through P.

Through G draw GN parallel to AM, and draw MN perpendicular to GN and join PN.

Then since PM, MN are perpendicular to AM, PN is also perpendicular to AM and therefore to GN.

Now $\qquad NG = MG \cos a = (g/\omega^2) \cos a,$
and $\qquad NM = MG \sin a = (g/\omega^2) \sin a.$

Hence NG is a fixed line in the same vertical plane as AM, and at a distance $(g/\omega^2) \sin a$ from it.

Also as the subnormal NG is constant $(g/\omega^2) \cos a$, the surface of equi-pressure must be generated by the revolution of a parabola of latus rectum $(2g/\omega^2) \cos a$, about its axis NG.

It should be noted that though this surface is fixed in *space*, it is not fixed relatively to the liquid, so that each element of the liquid is continually crossing a surface of equi-pressure.

EXAMPLES. CHAPTER III.

1. A fine tube bent into the form of an ellipse is held with its plane vertical, and is filled with n fluids whose densities are $\rho_1, \rho_2 \ldots \rho_n$ taken in order round the elliptic tube. If $r_1, r_2 \ldots r_n$ be the distances of the points of separation from either focus, prove that
$$r_1(\rho_1 - \rho_2) + r_2(\rho_2 - \rho_3) + \ldots r_n(\rho_n - \rho_1) = 0.$$
What is the corresponding theorem if the fluids do not fill the tube? [Pet., 1889.]

2. If there be n fluids arranged in strata of equal thickness and the density of the uppermost be ρ, of the next 2ρ, and so on, that of the last being $n\rho$: find the pressure at the lowest point of the nth stratum, and thence prove that the pressure at any point within a fluid whose density varies as the depth is proportional to the square of the depth. [M. T., 1854.]

3. A solid triangular prism, the faces of which include angles a, β, γ, is placed in any position entirely within a liquid; if $P, Q, R,$ be the thrusts on the three faces respectively opposite to the angles $a, \beta, \gamma,$ prove that
$$P \cosec a + Q \cosec \beta + R \cosec \gamma$$
is invariable so long as the depth of the centre of gravity of the prism is unchanged. [M. T., 1857.]

4. If the sea be supposed of an uniform temperature at which the compressibility per atmosphere is ·00004, the height of the water barometer being 32 feet; shew that the compression at a depth of two miles is about 13 in 1000, and that the reduction in that depth of water is about 70 feet. [Clare Coll., 1885.]

5. Six equal uniform tubes of small bore are united so as to form the edges of a regular tetrahedron, and so as to communicate freely with one another at the angular points. The tubes are filled with equal volumes of three heavy fluids of different densities,

and the tetrahedron is placed with one face horizontal and its vertex downwards; determine the position of equilibrium when each fluid is continuous and separated from the other two by smooth small pistons. [M. T., 1876.]

6. A circular tube of radius a (large in comparison with the bore) contains liquid filling 1/12 its circumference, and turns about the vertical diameter with uniform angular velocity. Prove that if the highest point of the liquid is in the horizontal diameter, the angular velocity is $2\sqrt{g/a}$. [M. T., 1874.]

7. A circular tube of small section is half full of liquid and in the surface at each side floats one of two small equal spheres which just fit the tube. The tube rotates with angular velocity ω about a vertical diameter: find the pressure at any point and shew that for values of $\omega >$ a certain quantity, the pressure is a maximum at a depth $g\omega^{-2}$ below the centre. [M. T., 1882.]

8. A circular tube of uniform bore, whose plane is vertical, contains columns of two liquids, whose respective densities are ρ, ρ', the respective columns subtending angles 2θ, $2\theta'$ at the centre of the circle. If a be the angle which the portion of the tube intercepted between the lowest point and the common surface of the liquids subtends at the centre of the tube, prove that

$$\rho \sin \theta \sin (\theta \pm a) = \rho' \sin \theta' \sin (\theta' \mp a).$$ [M. T., 1873.]

9. A box containing fluid is projected up a rough inclined plane, the angle of inclination being greater than the angle of friction. Shew that the free surfaces of the fluid in its position of rest relative to the box when going up and coming down are planes inclined to one another at an angle equal to twice the angle of friction for the box and the inclined plane. [Peterhouse, 1890.]

10. A railway train, travelling with a given acceleration, arrives at an incline, and after ascending to a ridge, descends at the same incline on the other side. Assuming that the pull of the engine and the resistance are the same throughout, determine the levels of the water surface in the boiler in going up and down the incline, and prove that the difference of the levels is equal to the angle between the inclines. [M. T., 1885.]

58 GENERAL THEOREMS RELATING TO PRESSURE.

11. Two smooth inclined planes of equal altitude are fixed back to back, and boxes containing liquid slide on the planes under gravity, the boxes being connected by a fine string passing over a pulley at the vertex of the planes. Prove that the free surfaces of the liquids will be parallel and equally inclined to the planes, if the weights of the boxes and the liquids they contain be proportional to the cosecants of the angles the planes make with the vertical. [M. T., 1883.]

12. A vessel contains n different fluids resting in horizontal layers and of densities $\rho_1, \rho_2 \ldots \rho_n$ respectively, starting from the highest fluid. A triangle is held with its base in the upper surface of the highest fluid, and with its vertex in the nth fluid. Prove that, if Δ be the area of the triangle and $h_1, h_2 \ldots h_n$ be the depths of the vertex below the upper surfaces of the 1st, 2nd, ... nth fluids respectively, the thrust on the triangle is

$$\frac{1}{3} \cdot \frac{g\Delta}{h_1^2} \{\rho_1(h_1^3 - h_2^3) + \rho_2(h_2^3 - h_3^3) + \ldots \rho_n h_n^3\}.$$

[Trinity Coll., 1889.]

13. The sides of a rectangle are in the ratio $\pi : 4$, and semicircles are described on the longer sides as diameters. Prove that, if the rectangle be immersed in water, with one of the shorter sides in the surface, the thrusts on the two parts external to both semicircles will together be equal to that on the part common to them.

[M. T., 1861.]

14. A cylinder, radius a, and height h, contains liquid to a height l. Shew that if the cylinder and liquid rotate about their common axis, which is vertical, the angular velocity in order that the liquid may just not flow out is

$$\frac{2}{a}\sqrt{g(h-l)} \quad \text{or} \quad \frac{2h}{a}\sqrt{\frac{g}{h+2l}},$$

according as l is $>$ or $< \frac{1}{2}h$. [Clare Coll., 1888.]

15. If liquid be contained in a cylinder whose axis is vertical, and which rotates with the liquid about a parallel axis, prove that the free surface will meet the surface of the cylinder in an ellipse.

[M. T., 1880.]

GENERAL THEOREMS RELATING TO PRESSURE. 59

16. Fluid is contained within a circular tube of radius a in a vertical plane which can rotate about a vertical axis. If the fluid subtend an angle θ at the centre, the least angular velocity to make the fluid divide into two parts is

$$\sqrt{(g/a)} \sec \tfrac{1}{4}\theta. \qquad \text{[Peterhouse, 1888.]}$$

17. A hollow cone vertex upwards, is three-quarters full of water and is set rotating about its axis which is vertical with an angular velocity equal to $\sqrt{8g/3h}\cot a$, where a is the semivertical angle and h the height of the cone. Prove that the thrust on the base is to the weight of the water in the vessel as 10 : 3.

[Peterhouse, 1889.]

18. Two hollow cones, filled with water, are connected together by a string attached to their vertices, which passes over a fixed pulley: prove that, during the motion, if the weights of the cones be neglected, the thrusts on their bases will be always equal, whatever be the forms and dimensions of the cones. If the heights of the cones be h, h', and heights mh, nh' be unoccupied by water, the thrusts on the bases during the motion will always be in the ratio

$$n^2 + n + 1 : m^2 + m + 1. \qquad \text{[M. T., 1861.]}$$

19. Two equal vertical cylinders of length l stand side by side and there is a free communication between their bases. Quantities of two fluids of densities ρ_1, ρ_3, which would fill lengths a and c respectively of the cylinders are poured in and rest in equilibrium, each fluid being continuous. A given quantity of a fluid of density ρ_2, intermediate between ρ_1 and ρ_3 is poured slowly into one of the cylinders. Find the position of equilibrium, noticing the different cases which may occur, and shew that, if the fluid reach the top of both cylinders at the same time, either

$$(\rho_1 - \rho_2)(2l - a - c) = c(\rho_1 - \rho_2) \quad \text{or} \quad a(\rho_1 - \rho_2) = c(\rho_2 - \rho_3).$$

[M. T., 1882.]

20. When a cylinder open at the top and half full of liquid revolves with angular velocity Ω about its axis, which is vertical, the liquid just reaches the upper rim; shew that the angular velocity in order that $1/n$th of the fluid may remain in the cylinder is $\Omega\sqrt{n}$.

[Clare Coll., 1887.]

21. A fine bent tube, on a vertical plane, has its branches AB, BC, inclined to the vertical at angles a, β on opposite sides, and contains two fluids which fill lengths a, b of the tube, the fluids meeting in the branch AB at a distance c from B: prove that they will meet at B, if the tube be turned in its own plane through an angle whose cosecant is

$$b/c \cdot \sin \tfrac{1}{2}(a+\beta) \sec \tfrac{1}{2}(a-\beta) \sec a - \tan a. \quad [\text{M. T., 1871.}]$$

22. A thin bent tube, open at both ends and of uniform bore, the two parts of which include a right angle, contains fluid, which, attracted towards the angle by a force $\mu \times$ the distance, is at rest relatively to the tube, while one part of the tube remains stationary and the other revolves with a uniform angular velocity ω: find the respective lengths of the tube occupied by the fluid.

23. A conical vessel without weight just filled with homogeneous incompressible fluid is attached to a fixed point by an elastic string attached to the vertex of the cone, and oscillates between positions in which the string is of its natural length and of twice its natural length respectively: find the time of an oscillation and the greatest pressure of the fluid on the base of the cone.

[M. T., 1869.]

24. Assuming that a mass of liquid contained in a vertical cylinder can rotate about the axis of the cylinder under the action of gravity only, in such a manner that the velocity at any point of the liquid varies inversely as the angular velocity of its distance from the axis of the cylinder: find the form and position of the free surface. [M. T., 1872.]

25. A hollow sphere filled with water has a small aperture in its diametral plane closed by an extensible membrane, and is spun round a vertical axis: determine in terms of the radius, the angular velocity necessary for a vacuum to be produced inside by bulging of the membrane in opposition to the atmospheric pressure. If the barometer stands at 30 inches and the radius is 3 inches, prove that the sphere must rotate about $29\tfrac{1}{2}$ times per second.

[M. T., 1888.]

GENERAL THEOREMS RELATING TO PRESSURE. 61

26. A semicircular tube has its bounding diameter horizontal, and contains equal volumes of n fluids of densities successively equal to $\rho, 2\rho, 3\rho, \ldots$ arranged in this order. Shew that if each fluid subtends an angle $2a$ at the centre, and the tube just holds them all, then
$$\tan na = (2n+1)\tan a. \qquad \text{[M. T., 1891.]}$$

27. If a vessel be in the shape of a paraboloid of revolution with its axis vertical, shew that no angular velocity will enable the vessel to contain any water, if a small hole be made at its lowest point.
[Clare Coll., 1884.]

28. A cone stands with its base on a horizontal table and is filled with fluid of varying density. If it is then turned so that the vertex is downwards, find the relation between the new and old positions of the different particles when equilibrium has been established in the new position. [Pet., 1886.]

29. A semicircle is immersed vertically in liquid with the diameter in the surface: shew how to divide it into any number of sectors, such that the thrust on each is the same.
[M. T., 1878.]

30. Shew that if liquid of density σ is rotating as if rigid with uniform angular velocity about a vertical axis, while the air above it is at rest, the free surface will be a paraboloid of revolution whose latus rectum is to that of a surface of equi-pressure in the liquid as $\sigma - \rho : \sigma$, if the variations in the density, ρ, of the air be neglected.
[Jesus Coll., 1891.]

31. Prove that if a mass of homogeneous fluid rotate about an axis and be acted on by a force to a point in the axis, varying inversely as the square of the distance, the curvatures of the meridian curve of the free surface at the equator and pole are respectively $1/a(1-m)$ and $(1-mb^3/a^3)/b$, where a and b are the equatorial and polar radii, and m is the ratio of the centrifugal force at the equator to the attraction there.

32. A closed vessel whose form is that of a right circular cylinder, of radius a with plane ends, perpendicular to the axis

of the cylinder, is nearly filled with liquid of density ρ, and placed with its axis vertical: a repelling centre of force ($\mu \times$ distance) resides at the centre of the base. Shew that if h, the height of the cylinder, be g/μ, the resultant thrusts on the top and on the bottom will be in the ratio

$$a^3 \;:\; a^3 + 2ah^2.\qquad \text{[Clare Coll., 1891.]}$$

33. Liquid is placed in a vessel and is subjected to the action of a force whose direction is perpendicular to the bottom of the vessel and which is directly proportional to the perpendicular distance from the bottom. Prove that the thrust on a rectangular area, which is vertically immersed in the liquid with one side resting on the bottom is $\tfrac{2}{3}pA$, where A is the area of the rectangle, and p is the pressure at the bottom, the upper side of the rectangle being in the free surface of the liquid. [Pet., 1891.]

CHAPTER IV.

Centre of Pressure.

40. Def. *The centre of pressure of any plane area exposed to fluid pressure is the point of it at which the resultant thrust on one side of the area acts.*

Since the action of the fluid is everywhere perpendicular to the area, the centre of pressure is the centre of a number of like parallel forces.

41. Prop. *The centre of pressure of a plane area inclined to the vertical in a homogeneous liquid is in a vertical line with, and at double the depth of, the centre of mass of the volume enclosed by the area, the effective surface of the liquid and vertical lines drawn through the perimeter of the area.*

If the area be divided into an infinite number of indefinitely small portions, the pressure on any one is proportional to the weight of the column of liquid standing on it (Art. 23). Also the small area is at double the depth of the centre of mass of the column. Hence the centre of pressure must be in a vertical line with the centre of mass of the volume, but at double the depth.

42. Prop. *If an area in a homogeneous liquid be turned about the line in which its plane meets the effective surface, the position of the centre of pressure in the area is unaltered.*

As the area turns, the depths of the different points below the effective surface alter in the same proportion, and therefore the pressures at the different points also alter in the same proportion. The centre of pressure will therefore not alter its position.

43. Prop. *Having given the centre of pressure of an area for one position, to determine it when the area is lowered a given depth, the liquid being homogeneous.*

Let P be the resultant thrust on the area A in the given position, as determined by Art. 35, C the corresponding centre of pressure; let z be the depth through which the area is lowered, ρ the density of the liquid.

By lowering the area through a distance z the pressure at every point of it is increased by $g\rho z$, and consequently the increase in the resultant thrust is $Ag\rho z$, aud acts at G, the centre of mass of the area.

The new centre of pressure then is the point where the resultant of P at C and $Ag\rho z$ at G acts.

Cor. We can by the above deduce the position of the centre of pressure when there is an atmospheric pressure, if we know it when there is no atmosphere.

For by Art. 23, the effect of introducing an atmospheric pressure Π is equivalent to increasing the depth by $\Pi/g\rho$.

44. Prop. *To find the centre of pressure of a parallelogram, the upper side of which is in the surface of a homogeneous liquid, not exposed to pressure.*

Let $ABCD$ be the parallelogram, AB being in the free surface. Join E, the middle point of AB, with F the middle point of CD.

CENTRE OF PRESSURE. 65

Divide the parallelogram into an infinite number of indefinitely small equal strips by drawing lines parallel to AB.

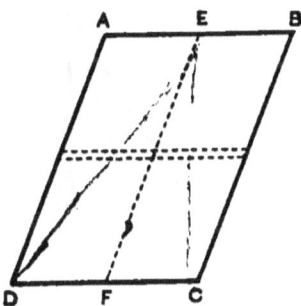

The thrust on any strip is proportional to the area of the strip and to its depth below the surface (Art. 23), and clearly acts at the centre of the strip, i.e. at a point in EF.

But the areas of the strips are all the same, and their depths are proportional to their distances from AB measured along EF.

We have therefore to find the centre of an infinite number of parallel forces acting at equal intervals along EF, each proportional to its distance from E. This must be the centre of mass of the triangle EDC.

Hence the centre of pressure is in EF at a distance from $E = \frac{2}{3} EF$.

45. Prop. *To find the centre of pressure of a triangle whose base is horizontal, and vertex in the surface of a homogeneous liquid, not exposed to pressure.*

Let A be the vertex of the triangle, BC its base. Join A with D the middle point of BC.

Divide the triangle into an infinite number of indefinitely small strips of equal breadth by drawing lines parallel to BC.

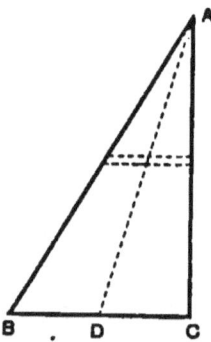

The thrust on each strip is proportional to its area and to its depth below A, and clearly acts at a point in AD.

The area of a strip is proportional to its distance from A measured along AD, as is also its depth.

Hence the thrust on each strip is proportional to the square of its distance from A.

We have therefore to find the centre of an infinite number of parallel forces acting at equal intervals along AD, each being proportional to the square of its distance from A.

But this is the centre of mass of a pyramid whose vertex is A, and the centre of mass of whose base is D.

Hence the centre of pressure of ABC is in AD at a distance $\frac{3}{4}AD$ from A.

46. Prop. *To find the centre of pressure of a triangle, whose base is in the surface of a homogeneous liquid, not exposed to pressure.*

CENTRE OF PRESSURE.

Let ABC be the triangle, BC being the base. Join A with D the middle point of BC.

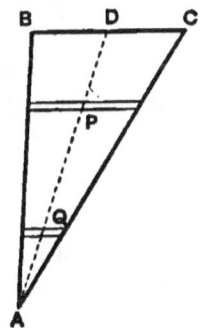

Divide the triangle into an infinite number of indefinitely small strips of equal breadth by drawing lines parallel to BC.

The centre of pressure of each strip is clearly in AD.

Consider two strips such that the distance of one from D is equal to that of the other from A. Let P, Q be their respective centres of pressure.

The thrust on the upper strip is proportional to its area and its depth conjointly, i.e. to $AP.PD$.

Similarly that on the lower strip is proportional to $DQ.QA$, i.e. is equal to that on the upper.

The centre of pressure of this pair as well as of every similar pair is therefore half-way between A and D.

Hence the centre of pressure of the triangle is the middle point of AD.

COR. *The thrust on a triangle whose base is in the surface of a homogeneous liquid is equivalent to two equal forces acting at the middle points of the two sides.*

EXAMPLES.

1. Find the depth of the centre of pressure of a trapezium, (1) with one of the parallel sides, (2) with one of the other sides, in the surface.

2. Shew that the depth of the centre of pressure of a rhombus totally immersed with one diagonal vertical and its centre at a depth h is $(\frac{1}{12}a^2 + h^2)/h$, where a is the length of the vertical diagonal.
[Jesus Coll., 1889.]

3. A triangle is wholly immersed in a liquid with its base in the surface. Prove that a horizontal straight line drawn through the centre of pressure of the triangle divides it into two portions, the thrusts on which are equal. [Pet. 1891.]

4. In Ex. 3, p. 28, find the centre of pressure on a parallelogram, with one side in the surface of the oil, and the other in the mercury surface, assuming that there is no atmospheric pressure.

5. Given that the centre of pressure of a circular disc of radius r with one point in the surface is at a distance p from the centre, shew that for a disc of radius R wholly immersed with its centre at a distance h from the surface, the distance between the centre of the circle and the centre of pressure is pR^2/hr. [M. T. 1881.]

6. Shew that the centre of pressure of a parallelogram immersed in a liquid with one angular point in the surface and one diagonal horizontal, lies in the other diagonal and is at a depth equal to $\frac{7}{12}$ of the depth of its lowest point. [Pet. 1888.]

47. Prop. Centre of pressure of any triangle. *To find the Centre of Pressure of any triangle wholly immersed in a homogeneous liquid.*

Let ABC be the triangle, D, E, F being the middle points of the sides BC, CA, AB respectively: let α, β, γ be the distances of A, B, C respectively from the horizontal line NK, where the plane of the triangle meets the effective surface.

CENTRE OF PRESSURE. 69

Let Δ be the area of the triangle, and suppose A to be the highest point.

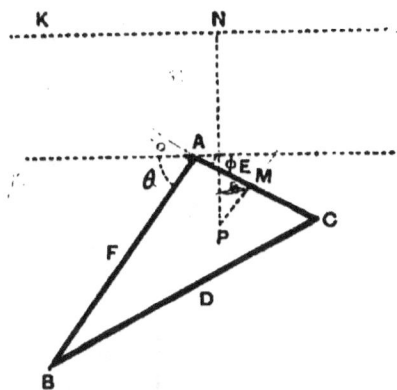

Let P be any point in the triangle. From P draw PM parallel to AB meeting AC in M, and PN perpendicular to NK.

Then by projecting on PN, we see that
$$PN = a + PM \sin \theta + MA \sin \phi,$$
where θ, ϕ are the angles which AB, AC make with NK.

Let $PN = z$, $PM = x$, $MA = y$,
$$\therefore z = a + x \sin \theta + y \sin \phi.$$

The pressure at $P = \mu z$, where μ is constant (Art. 23).

Thus the pressure can be divided into three parts, μa, $\mu x \sin \theta$ and $\mu y \sin \phi$.

The thrust due to the first part is $\mu \Delta a$, and acts at the C.M. of the triangle. This may be replaced by $\tfrac{1}{3} \mu \Delta a$, acting at each of the points D, E, F.

The thrust due to the second part of the pressure is (Art. 35) $\tfrac{1}{2} \mu \Delta c \sin \theta$, and (Cor. Art. 46) is equivalent to $\tfrac{1}{8} \mu \Delta c \sin \theta$ at D and at F. Similarly the thrust due to the third part of the pressure is $\frac{1}{3}\mu\Delta b \sin\phi$, and is equivalent to $\frac{1}{6}\mu\Delta b \sin\phi$ at D and at E. Hence the whole thrust is equivalent to

$\frac{1}{3}\mu\Delta (\alpha + \frac{1}{2}c \sin\theta)$ at F, $\frac{1}{3}\mu\Delta (\alpha + \frac{1}{2} b \sin\phi)$ at E,

and $\frac{1}{3}\mu\Delta (\alpha + \frac{1}{2} c \sin\theta + \frac{1}{2}b \sin\phi)$ at D,

i.e. to forces at D, E, F proportional to their respective distances from NK, i.e. proportional to their depths.

This determines the position of the centre of pressure. Its distance from NK

$$= \frac{\{\frac{1}{2}(\alpha+\beta)\}^2 + \{\frac{1}{2}(\beta+\gamma)\}^2 + \{\frac{1}{2}(\gamma+\alpha)\}^2}{\frac{1}{2}(\alpha+\beta) + \frac{1}{2}(\beta+\gamma) + \frac{1}{2}(\gamma+\alpha)}$$

$$= \frac{\alpha^2 + \beta^2 + \gamma^2 + \beta\gamma + \alpha\gamma + \alpha\beta}{2(\alpha+\beta+\gamma)}.$$

It is easy to deduce that the thrust on the triangle is equivalent to forces proportional to $2\alpha+\beta+\gamma$, $\alpha+2\beta+\gamma$, $\alpha+\beta+2\gamma$, at A, B, C respectively.

48. PROP. *To find the centre of pressure of a circle, wholly immersed with its plane vertical in a homogeneous liquid.*

Let AB be the vertical diameter of the circle, O its centre and r its radius. Let ρ be the density of the liquid; let h be the depth of O below the effective surface.

Construct a hemisphere on the circle as base, and consider the equilibrium of the liquid contained in it.

The forces acting on this liquid are

(i) its weight $\frac{2}{3}\pi r^3 g\rho$, vertically downwards through the C.M. G, which is situate in the radius at right angles to the circle at a distance $\frac{3}{8}r$ from O;

CENTRE OF PRESSURE. 71

(ii) the thrust on the circle $\pi r^2 h g \rho$ through the centre of pressure C, which is clearly in the line OB;

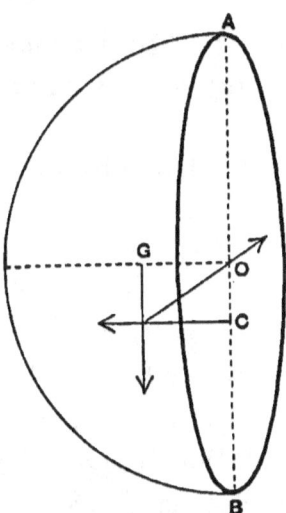

(iii) the resultant thrust on the curved surface, which must act through O.

Taking moments about O, we obtain

$$\tfrac{2}{3}\pi r^3 g \rho \times OG = \pi r^2 h g \rho \times OC;$$

$$\therefore OC = \frac{2}{3} \times \frac{3}{8} \cdot \frac{r^2}{h} = \frac{r^2}{4h}.$$

COR. If the plane be *not* vertical, we shall obtain (Art. 42) the same expression for OC, provided h denote the distance of the centre from the effective surface measured in the plane of the circle.

The centre of pressure of a circle being known, we can deduce that of an *ellipse*; it will clearly be in the diameter conjugate to the horizontal diameter, and at a distance from the centre $a^2/4h$, where a is the semi-diameter, and h the distance of the centre from the effective surface, measured along the diameter.

49. It is obvious that as any area moves down to an infinite depth, the centre of pressure approaches and ultimately coincides with the centre of mass, since the pressure over the area becomes ultimately uniform.

50. The following method is applicable to finding the centre of pressure of any plane area immersed in a homogeneous liquid.

Let the area be divided into an infinite number of indefinitely small portions.

Let α be any one of these portions, z its depth below the effective surface, ρ the density of the liquid. Then the thrust on this small portion is $g\rho z\alpha$, and the depth of the centre of pressure below the effective surface is

$$\Sigma(g\rho z^2\alpha)/\Sigma(g\rho z\alpha), \text{ or } \Sigma(z^2\alpha)/\Sigma(z\alpha).$$

With the notation of the Integral Calculus, the depth below the effective surface is $\iint z^2 dS / \iint z\, dS$, where dS is the indefinitely small area at depth z.

In the more general case, where the pressure is not necessarily proportional to the depth below the effective surface, the depth of the Centre of Pressure is $\iint zp\, dS / \iint p\, dS$, where p is the pressure at the depth z.

ILLUSTRATIVE EXAMPLES.

1. *If a quadrilateral lamina ABCD in which AB is parallel to CD be immersed in liquid with the side AB in the surface, the centre of pressure will be at the point of intersection of AC and BD if $AB^2 = 3CD^2$.*

[M. T. 1875.]

Let $AB = a$, $CD = b$, and let h be the depth of CD.

By Art. 47, we may replace the thrust on a triangle whose angular points are at depths α, β, γ respectively, by forces at those points, proportional to the products of the area of the triangle and the magnitudes $2\alpha + \beta + \gamma$, $\alpha + 2\beta + \gamma$, $\alpha + \beta + 2\gamma$, respectively.

Applying this to each of the two triangles ABD, CBD, whose areas are proportional to a, b respectively, we find that the thrust on ABD is equivalent to forces, μah at A, μah at B, and $2\mu ah$ at D. Similarly

CENTRE OF PRESSURE.

the thrust on BDC is equivalent to $2\mu bh$ at B, $3\mu bh$ at C, and $3\mu bh$ at D.

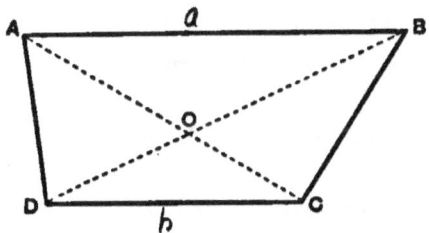

Hence the whole thrust is equivalent to μah at A, $\mu(a+2b)h$ at B, $3\mu bh$ at C, and $\mu(3b+2a)h$ at D.

The resultant of the forces at A and C will act at O, the intersection of AC and BD,

if $$\mu ah \cdot AO = 3\mu bh \cdot OC,$$
if $$a^2 = 3b^2;$$
$$\therefore AO/OC = a/b.$$

Similarly the resultant of the forces at B and D acts at O.

2. *If a quadrilateral area be entirely immersed in water, and a, β γ, δ, be the depths of its four corners, and h that of its centre of gravity, shew that the depth of its centre of pressure is*

$$\frac{1}{2}(a+\beta+\gamma+\delta) - \frac{1}{6h}(\beta\gamma+\gamma a+a\beta+a\delta+\beta\delta+\gamma\delta).$$

[M. T. 1881.]

Join AC: let S_1, S_2 be the areas of the triangles ABC, ADC re-

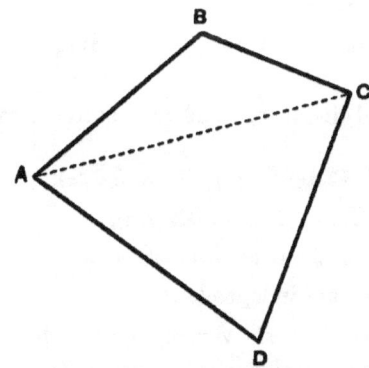

spectively. The depths of their centres of gravity are
$\frac{1}{3}(a+\beta+\gamma)$, and $\frac{1}{3}(a+\delta+\gamma)$ respectively.

CENTRE OF PRESSURE.

Since h is the depth of the centre of gravity of the quadrilateral

$$h(S_1+S_2) = \tfrac{1}{3}(\alpha+\beta+\gamma)S_1 + \tfrac{1}{3}(\alpha+\delta+\gamma)S_2 \dots\dots\dots\dots\dots(1).$$

This equation determines the ratio of $S_1 : S_2$.

The depths of the centres of pressure of the triangles ABC, ADC respectively, are (Art. 47)

$$\frac{1}{2} \cdot \frac{\alpha^2+\beta^2+\gamma^2+\alpha\beta+\alpha\gamma+\beta\gamma}{\alpha+\beta+\gamma}, \quad \frac{1}{2} \cdot \frac{\alpha^2+\delta^2+\gamma^2+\alpha\delta+\alpha\gamma+\delta\gamma}{\alpha+\gamma+\delta};$$

and the thrusts on them are proportional to

$$\tfrac{1}{3}S_1(\alpha+\beta+\gamma), \quad \tfrac{1}{3}S_2(\alpha+\delta+\gamma) \text{ respectively};$$

∴ z, the depth of the centre of pressure of the quadrilateral is given by

$$zh(S_1+S_2) = \tfrac{1}{6}S_1(\alpha^2+\beta^2+\gamma^2+\alpha\beta+\alpha\gamma+\beta\gamma)$$
$$+ \tfrac{1}{6}S_2(\alpha^2+\gamma^2+\delta^2+\alpha\gamma+\alpha\delta+\gamma\delta);$$

substituting for S_1/S_2 from (1), we obtain the required result.

3. *Shew that, if a lamina totally immersed in a homogeneous liquid be moved without rotation, the vertical distance between the centre of pressure and the centre of gravity varies inversely as the depth of the centre of gravity.*

If the fluid is heterogeneous with a slowly varying density, the vertical distance between the centre of gravity and the centre of pressure will, at considerable depths, be proportional to the quotient of the density by the pressure at the centre of gravity. [M. T. 1880.]

Let h be the depth of the centre of gravity, G, of the area; let z be the distance, measured vertically downwards, of any indefinitely small area, a, below G. Then $\Sigma(az) = 0$.

The thrust on the area $a = g\rho(h+z)a$, if ρ is the density of the liquid.

Hence the vertical distance, $\bar{z}$, of the centre of pressure below G is given by

$$\bar{z}\Sigma\overline{(g\rho h+za)} = \Sigma(g\rho z\overline{h+za}),$$
$$\therefore \quad \bar{z}h\Sigma(a) = \Sigma(z^2 a),$$
$$\therefore \quad \bar{z} \text{ varies inversely as } h.$$

∵ $\Sigma(a)$ and $\Sigma(az^2)$ are independent of h.

When the liquid is of slowly varying density, let p be the pressure at G, when it is at a great depth h.

Since the dimensions of the area are small compared with h, we may suppose the density over the area to be constant and the same as at G, ρ.

CENTRE OF PRESSURE. 75

Hence the thrust on the area a, distant z vertically from G,

$$= a\,(p + g\rho z),$$

and the vertical distance of the centre of pressure from G

$$= \frac{\Sigma a\,(p + g\rho z)\,z}{\Sigma a\,(p + g\rho z)}$$

$$= \frac{g\rho \Sigma\,(az^2)}{p\Sigma\,(a)},$$

since $\Sigma\,(az) = 0$;

i.e. the vertical distance varies as ρ/p.

EXAMPLES. CHAPTER IV.

1. A square is just immersed vertically in heavy fluid with one corner in the surface and a side inclined at an angle θ to the vertical: prove that the distances of the centre of pressure from the two sides of the square which meet in the surface are respectively

$$\frac{a}{6} \cdot \frac{4\sin\theta + 3\cos\theta}{\sin\theta + \cos\theta} \quad \text{and} \quad \frac{a}{6} \cdot \frac{4\cos\theta + 3\sin\theta}{\sin\theta + \cos\theta},$$

where a is the length of a side. [Jesus Coll., 1890.]

2. ACB is a triangle immersed in a liquid. The side AB is in the surface and is divided in D so that $6AD.DB = AB^2$. Lines DE, DF drawn parallel to AC and BC form the parallelogram $DECF$. Prove that the depths of the centres of pressure of $DECF$ and ACB are in the ratio 11 : 9. [Clare Coll., 1890.]

3. A lamina in the form of a right-angled triangle is just immersed in a fluid and the centre of pressure is vertically below the right angle. Shew that the tangent of twice the angle which one of the sides makes with the hypothenuse is double of the tangent of twice the angle which it makes with the surface of the fluid.
[Jesus Coll., 1875.]

4. An isosceles triangle of vertical angle $2a$, is placed vertically with its vertex in the surface of a liquid and its base inclined at an angle θ to the horizon. Shew that the depth of its centre of pressure is

$$l \cdot \frac{3\cos^2\theta\cos^2 a + \sin^2\theta\sin^2 a}{4\cos\theta\cos a};$$

l being one of the equal sides. [Clare Coll., 1888.]

5. A regular hexagon is immersed in homogeneous liquid with one side in the surface, prove that the depth of its centre of pressure is to that of its centre of mass as 23 to 18.

6. A rectangle is immersed in n fluids of densities $\rho, 2\rho, 3\rho...n\rho$; the top of the rectangle being in the surface of the first fluid and the area immersed in each fluid being the same: shew that the depth of the centre of pressure of the rectangle is $\dfrac{3n+1}{2n+1} \cdot \dfrac{h}{2}$, where h is the depth of the lower side. [Trin. Coll., 1891.]

7. A parallelogram, whose plane is vertical and centre at a depth h below the surface, is totally immersed in a homogeneous fluid. Shew that, if a, b be the lengths of the projections of the sides on a vertical line, the depth of its centre of pressure will be

$$h+(a^2+b^2)/12h.$$ [M. T., 1882.]

8. A cubical box filled with water is closed by a lid without weight which can turn freely about one edge of the cube, and a string is tied symmetrically round the box in a plane which bisects the edge; shew that, if the lid be in a vertical plane with this edge uppermost, the tension of the string is one-third of the weight of the water. [Jesus Coll., 1886.]

9. The embankment of a reservoir is composed of thin horizontal rough slabs of stone of density ρ and whose coefficient of friction is μ. The top of the embankment is a feet wide, the side in contact with the water is vertical, and na feet deep. Shew that the slope of the outer side to the horizon must be < the smaller of the angles

$$\cot^{-1}(1/\mu\rho - 2/n), \quad \cot^{-1}\{\sqrt{(1/2\rho + 3/4n^2)} - 3/2n\}.$$

[M. T., 1882.]

10. A plane area is completely immersed in water, its plane being vertical; it is made to descend in a vertical plane without any rotation and with uniform velocity, shew that the centre of pressure approaches the horizontal through its centre of mass with a velocity which is inversely proportional to the square of the depth of its centre of mass. [Peterhouse, 1888.]

11. Prove that the depth of the centre of pressure of a parallelogram, two of whose sides are horizontal and at depths h, k below the surface of a liquid whose density varies as the depth below the surface, is

$$\frac{3}{4} \cdot \frac{h^3 + h^2k + hk^2 + k^3}{h^2 + hk + k^2}.$$ [Jesus Coll., 1884.]

12. A portion of the side of a vessel, in the form of a vertical plane triangle of altitude a, with its vertex uppermost and base horizontal, is movable about a horizontal line in its own plane, whose height above the base is na ($4n$ being less than 1): prove that this portion will be in equilibrium if the vessel contains water to a depth

$$\{1 + n - (1 - 4n + n^2)^{\frac{1}{2}}\} a.$$ [M. T., 1873.]

13. A vessel in the form of a regular tetrahedron rests with one face on a horizontal table. The other faces are uniform plates, each of weight w, which can turn freely about their lowest edges, and when shut fit closely. Through a hole at the top water is poured in and the sides are pressed out when the depth of the water is m times the height of the vessel. Shew that if the weight of water poured in be pw, then

$$9p(2m^2 - n^3) = 2(m^2 - 3m + 3).$$ [M. T., 1890.]

14. An ellipse, semi-axes (a, b), is immersed in fluid in a vertical plane so that the axes make equal angles with the vertical. Shew that its centre of pressure lies in a line through the centre making with the vertical an angle

$$\tan^{-1}(a^2 - b^2)/(a^2 + b^2),$$

and at a depth $\quad h + (a^2 + b^2)/8h,$

h being the depth of the centre of the ellipse. [Pet., 1889.]

15. One asymptote of a hyperbola lies in the surface of a fluid; find the depth of the centre of pressure of the area included between the immersed asymptote, the curve and two given horizontal lines in the plane of the hyperbola.

16. A hyperbola is immersed in water with an asymptote AB in the surface and from any point C on the curve a line CA is drawn (1) touching the curve, (2) parallel to the other asymptote. Prove that the depth of the centre of pressure of the area between AB, AC and the curve is in the former case one-fourth, and in the latter case one-third the depth of C, and shew how to find its exact position. [M. T., 1877.]

17. Prove the following theorems for determining the centre of pressure of a vertical area:

(1) For similarly situated positions the locus of the centre of pressure in the area is a line through the centre of gravity and the distance from the centre of gravity is inversely as the magnitude of the pressure.

(2) If the centre of gravity G is fixed, and the centres of pressure when a given line in the area is horizontal and vertical are respectively C_1, C_2; then when the line is inclined at an angle θ to the horizontal, the centre of pressure is at C where GC meets C_1C_2 in D so that

$$C_1 D \cos \theta = C_2 D \sin \theta,$$

and $\qquad GC = GD(\sin \theta + \cos \theta).$ [M. T., 1888.]

18. A rectangle (sides $2a$, $2b$) is completely immersed in water rotating about a vertical axis: its plane is vertical and one side ($2b$) touches the free surface at its lowest point, the point of contact being the middle point of the side: the rectangle rotates with the same angular velocity $(ng/b)^{\frac{1}{2}}$ as the water. Shew that the depth of the centre of pressure of the rectangle is

$$a(8a + 6h + nb)/(6a + 6h + nb),$$

when h is the height of the water barometer.

19. A closed cubical box with two faces horizontal is just filled with liquid and made to rotate about a vertical axis through its centre with uniform angular velocity: one of the sides is attached by fastenings at its corners only: find the reactions at these points supposing that the reactions are all perpendicular to the side, and the reactions at the upper points are equal. Shew that if the reactions at the upper are to those at the lower points in the ratio of 2 to 3, then the latus rectum of a surface of equal pressure is equal to the length of an edge of the box. [M. T., 1891.]

CHAPTER V.

Floating Bodies.

51. Prop. *To find the resultant thrust on a solid wholly or partly immersed in a homogeneous liquid.*

Let us suppose the solid to be removed and the gap it made in the liquid to be filled up with some new liquid of the same kind. Now the pressure at every point of the surface of this introduced liquid is the same as it was on the corresponding point of the solid since it depends on the depth below the effective surface, and therefore the resultant thrusts on the new or displaced liquid and on the solid are the same.

But the displaced liquid is clearly in equilibrium under the action of its weight, acting vertically downwards through its centre of gravity, and the resultant thrust. The latter must therefore be equal and opposite to the former.

Hence **the resultant thrust on the solid is equal to the weight of the liquid displaced by the solid, and acts vertically upwards through the centre of gravity of this displaced liquid.**

This is known as **Archimedes' theorem.**

NOTE. It should be observed that it is assumed that the liquid can flow *all round* the immersed part of the solid: for instance, a stone lying on the base of the vessel containing the liquid but with no liquid below it, and a stick thrust through the side of the vessel below the level of the liquid, are not cases to which the proposition applies.

We can extend this proposition to the case of a solid immersed partly in one fluid and partly in another, air and water, for instance, and also to the case of one immersed in a number of different fluids, arranged in layers of different densities. In this case, however, we must be careful to remember that by the displaced fluid is meant the whole mass which fills up the gap made by the solid, when the gap in each layer is filled up by a new liquid of the corresponding density.

52. DEF. *The resultant thrust on a solid immersed in a fluid is termed the* **Force of Buoyancy,** *and the centre of gravity of the displaced fluid is termed the* **Centre of Buoyancy.**

53. PROP. *To find the resultant thrust on any surface, one side of which is exposed to liquid pressure.*

In Art. 35 we have seen how to determine the magnitude of the resultant thrust on any *plane* surface, and we have in Arts. 44—48, found the centre of pressure in the case of certain areas. In the more general case of a *curved* surface, when the thrusts on the different portions are not in the same direction, their equivalent is not of necessity *a single force:* we can however determine certain forces to which they are equivalent.

Let the surface be divided into an infinite number of indefinitely small portions and let the thrust on each be resolved in three directions, one vertical and the other two any given horizontal directions at right angles to one another. We shall now shew how the resultants of these resolved parts can be severally found.

FLOATING BODIES.

To determine the resultant vertical thrust.

Through the perimeter AB of the surface draw vertical lines to meet the effective surface in the curve ab.

The vertical thrust on any indefinitely small area PQ of the surface is equal to the weight of the vertical column

Fig. i.

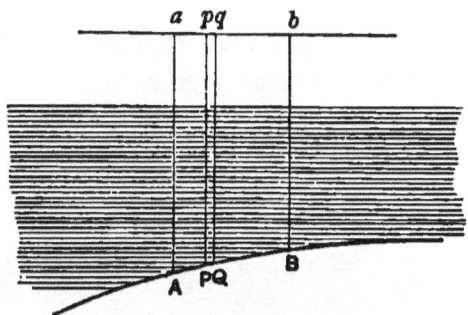

of liquid $PQqp$ which could stand on the area, its upper end pq being in the effective surface. The vertical thrust is *downwards*, if the actual liquid is *above PQ* as in fig. i.; and *upwards* if it is *below* as in fig. ii.

Fig. ii.

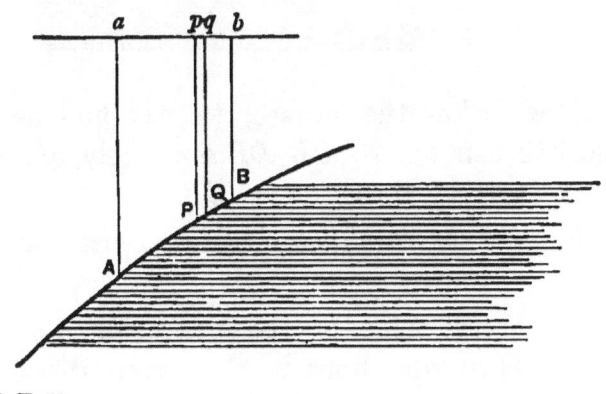

Hence if, as in fig. i., the liquid is at every point *above* the surface, the resultant vertical thrust on AB is *downwards* and equal to the weight of the liquid which would fill up the space $ABba$.

If, as in fig. ii., the liquid is at every point *below* the surface, the resultant vertical thrust is *upwards*, and equal to the weight of liquid which would fill up the space $ABba$.

In each case the vertical thrust acts through the centre of mass of the volume $ABba$.

If, as in fig. iii., the actual liquid is in some parts above

Fig. iii.

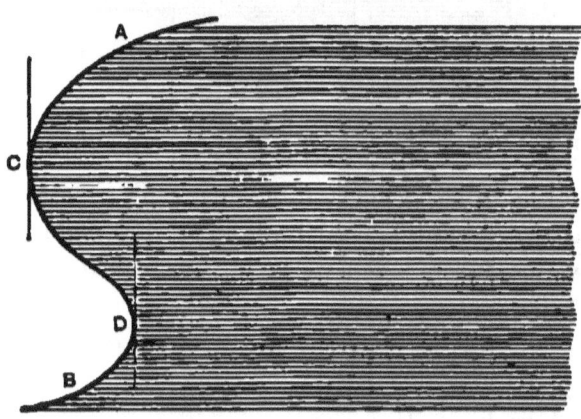

and in others below the surface, we can find as before the vertical thrusts on AC, CD, DB separately, and thence the resultant vertical thrust on $ACDB$.

To determine the resultant thrust in a given horizontal direction.

Through every point of the perimeter AB of the surface draw horizontal lines in the given direction to

FLOATING BODIES. 83

meet a vertical plane perpendicular to them in the closed curve *ab*.

Considering the equilibrium of the liquid filling up the space *ABba*, we see that the only forces acting on it

Fig. iv.

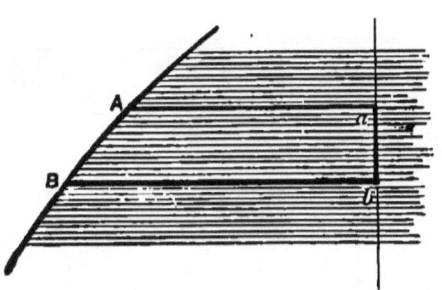

in the given direction are the thrust on the flat end *ab*, and the resultant thrust in the given direction on *AB*. These two must therefore be equal, i.e. the resultant thrust in the given direction is equal to the thrust on the projection of *AB* on a plane perpendicular to the given direction.

As this is true of every elementary portion of the surface, the line of action of this resultant thrust passes through the centre of pressure of *ab*.

Fig. v.

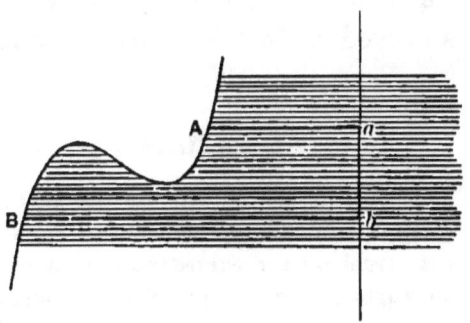

The above reasoning still applies when the surface is as in fig. v.

If the perimeter of the surface be a *plane* curve, the resultant thrust can be obtained in a simpler way.

Let ABC be a surface bounded by a plane AC. The

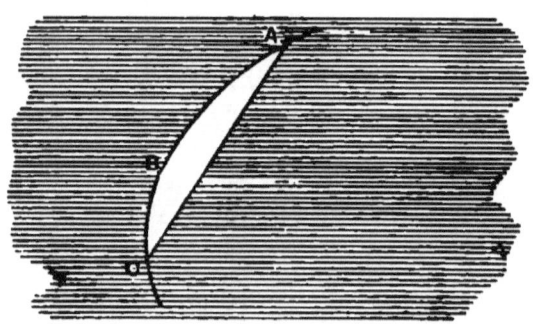

thrust on ABC will not be affected by supposing the part of the surface on the other side the plane AC to be removed and its place to be filled up with liquid. Then the resultant thrust on the body bounded by the curved surface ABC and the plane surface AC is the resultant of that on ABC and that on AC. The thrust on the whole body is obtained by the principle of Archimedes, and the thrust on the plane AC is obtained by the methods of previous chapters; these two being known the thrust on the curved surface is easily obtained.

EXAMPLES.

1. A solid hemisphere is placed with its base inclined to the surface of a liquid, in which it is just totally immersed, at a given angle α, shew that the resultant thrust on the curved portion of the surface will be equal to twice the weight of the liquid displaced if $\tan \alpha = 2$.

[Clare Coll., 1887.]

FLOATING BODIES.

2. A regular tetrahedron whose edges are of length a is completely immersed in water, with one of its faces horizontal, and the opposite vertex downwards. Having given the depth d of the horizontal face, find the resultant thrust, (1) on the tetrahedron, (2) on three of the faces including the horizontal one.

3. A hemispherical bowl is filled with water: shew that the resultant thrust on the bowl is π times the horizontal thrust on the portion on one side a vertical diametral plane.

4. A solid is formed by the revolution of a semicircle of radius a about its bounding diameter through an angle a, and the solid is immersed with one plane face in the surface of a liquid: prove that the magnitude of the resultant thrust on the curved surface of the solid is

$$\tfrac{2}{3} a^3 g \rho \left\{(a - \sin a \cos a)^2 + \sin^4 a\right\}^{\tfrac{1}{2}}. \quad \text{[Jesus Coll., 1890.]}$$

5. Find the resultant thrust on the curved surface of a right circular cone on an elliptic base, when placed with its axis vertical and vertex upwards and at a given depth below the surface of a uniform liquid.

6. A right circular cone is divided into two parts by a plane through its axis—one of these portions is just immersed vertex downwards in water. Shew that the resultant thrust on the curved surface of any frustum of the semi-cone acts in a direction making an angle $\tan^{-1}(\tfrac{1}{2}\pi \tan a)$ with the horizontal where a is the semi-vertical angle of the cone.

[Peterhouse, 1887.]

7. A conical wine-glass is filled with water and placed in an inverted position upon a table: shew that the resultant thrust of the water on the glass is two-thirds that on the table. [M. T., 1858.]

8. A vessel of water is placed in one scale of a balance, and there is a weight in the other which will just counterbalance it. Will the equilibrium be disturbed if a person dips his finger in the water without touching the sides of the vessel? Give reasons for your answer.

9. With what acceleration would a piece of cork (sp. gr. ·5) rise if plunged below the surface of some water and then released?

10. A solid cone is just immersed with a generating line in the surface: if θ be the inclination to the vertical of the resultant thrust on the curved surface; prove that

$$(1 - 3\sin^2 a)\tan \theta = 3 \sin a \cos a;$$

$2a$ being the vertical angle of the cone. [St John's Coll., 1881.]

FLOATING BODIES.

11. A closed cylinder the diameter of whose base is equal to its length is full of water and hangs freely by a string fastened to a point in its upper rim: prove that, the weight of the cylinder being neglected, the vertical and horizontal components of the resultant thrust on its curved surface are each half the weight of the water. [M. T., 1874.]

12. Find the vertical thrust on the lower half of the curved surface of a cylinder, immersed in any manner in a liquid, the dividing plane passing through the horizontal tangent lines at the highest and lowest points. [M. T., 1869.]

54. Prop. Body floating freely. *To find the conditions of equilibrium satisfied by a solid floating freely in a fluid.*

The forces acting on the solid are

(i) its weight, acting vertically downwards through its centre of gravity;

(ii) the force of buoyancy, which is equal the weight of the fluid displaced, and acts vertically upwards through the centre of buoyancy.

Hence these forces must be equal and in the same straight line, i.e. *the weight of the solid is equal to the weight of fluid displaced, and the centres of gravity and buoyancy are in the same vertical line.*

55. Prop. *When a solid of volume V and density ρ is floating in a liquid of density ρ', the volume immersed is $V\rho/\rho'$.*

Let V' be the volume immersed.

Then the weight of the liquid displaced $= g\rho'V'$, and the weight of the solid $= g\rho V$;

$$\therefore g\rho'V' = g\rho V \text{ (Art. 54),}$$

$$\therefore V' = V\rho/\rho'.$$

FLOATING BODIES. 87

Cor. Since V' cannot be greater than V, it follows that ρ' cannot be less than ρ, or a solid cannot float in a liquid of less density than its own. If the density of the liquid be less than that of the solid, the latter will sink; if the density of the solid be less than that of the liquid, the former will rise to the surface of the liquid until the volume immersed is that given by the formula.

NOTE. It should be quite clear what is meant by the density of the solid in the above formula. For instance a *hollow* iron ball may be capable of floating in water though the density of *iron* is much greater than that of water, because in calculating the density of the *ball*, we must take into account the space inside, which may be a vacuum or filled with air, so that the *mean density* of the ball may be less than that of water.

EXAMPLES.

1. A piece of iron weighing 275 grammes floats in mercury of density 13·6 with ⅔ of its volume immersed. Determine the volume and density of the iron.

2. A cylinder floats between two fluids with its axis vertical, its height being equal to the depth of the upper fluid: compare the thrusts on the two ends of the cylinder, the densities of the fluids and of the cylinder being given. [M. T., 1856.]

3. A right circular cylinder floats in water with its axis vertical, half its axis being immersed: assuming the specific gravity of air to be ·0013, find that of the cylinder. [Jesus Coll., 1880.]

4. A cylindrical vessel, the radius of the base of which is 1 foot, contains water; if a cubic foot of cork (sp. gr. = ·24) be allowed to float in the water, find the additional thrust sustained by the base.
[M. T., 1849.]

5. A steamer loading 30 tons to the inch near the water line in fresh water is found after a 10 days' voyage, burning 60 tons of coal a day, to have risen 2 feet in sea water at the end of the voyage: prove that the original displacement of the steamer was 5720 tons, taking a cubic foot of fresh water as 62·5 lbs. and of sea water as 64 lbs.

6. Prove that a homogeneous solid in the form of a right circular cone can float in a liquid of twice its own density with its axis horizontal. [St John's Coll., 1881.]

7. A heavy hollow right cone, closed by a base without weight, is totally immersed in a fluid: find the force which will sustain it with its axis horizontal.

8. A body is floating in water and a hollow vessel is inverted over it and depressed: what effect will be produced in the position of the body (1) with reference to the surface of the water within the vessel, (2) with reference to the surface of the water outside? [M. T., 1857.]

9. A square lamina is placed vertically in a fluid of double its density: prove that it can rest only with an edge or diagonal vertical. [M. T., 1867.]

10. Shew that a uniform lamina in the form of a parallelogram cannot float in a liquid so as to have two angular points at the same depth below the surface unless it be rectangular or equilateral.

11. One end of a thin rod of uniform section is made of a substance whose specific gravity is ·5, and the remainder of a substance whose specific gravity is 1·5: find the proportion of their lengths, in order that the rod may be able to float in an inclined position in water. [M. T., 1866.]

12. A conical vessel floats in water, with its vertex downwards and a certain depth of its axis immersed: when filled up to the depth originally immersed, it sinks till its mouth is on a level with the surface of the water. Find what portion of the axis was originally immersed.
[M. T., 1873.]

56. The Balloon. A balloon generally consists of a light nearly spherical silken envelope, capable of holding a large amount of gas. The gas may be any gas lighter than atmospheric air; hydrogen, coal gas, and air heated to a temperature above that outside are often used. To the balloon is attached a light car capable of holding one or more persons. When the balloon is filled, it will rise, provided the mean density of the whole including the car and the persons in it, is less than that of the air. The balloon will go on rising until it has reached the height, at which the atmospheric density is equal to that of the whole balloon.

***57.** We may extend the proposition of Art. 51 to the case of a body immersed in a fluid which is acted on by forces other than gravity, or immersed in relative equilibrium in a fluid in motion, as for instance one revolving about a vertical axis or moving with uniform acceleration.

In the former case, when the solid is removed, we must suppose that its place is taken by new fluid, which follows the law of density, necessary for the fluid to be in equilibrium. Then as before, since the pressure at every point of the solid is the same as that at the corresponding point of the displaced fluid, the resultant thrust on the solid is the same as that on the displaced fluid. Hence *when a fluid is at rest, the resultant thrust on any solid immersed in it is equal and opposite to the resultant external force on the displaced fluid.*

Hence as in Art. 54, when the solid is in equilibrium, the resultant external force on the solid must be equivalent to the resultant external force on the fluid displaced, since each counterbalances the same resultant thrust.

When the solid is in equilibrium *relative* to fluid in motion, the resultant thrust and the resultant external force must together give the solid its acceleration: the resultant thrust is obtained from the consideration that together with the resultant external force on the fluid displaced it would give the latter the acceleration which the solid has.

It should be observed that in these cases, the resultant external force, and the resultant thrust are not necessarily *single forces,* since the forces to which they are equivalent may not reduce to single forces.

Ex. 1. Find the conditions of equilibrium of a body floating in relative equilibrium in water contained in a vessel which is sliding under gravity down a smooth inclined plane.

Ex. 2. A piece of cork, of weight w, and specific gravity σ, is kept totally immersed in a vessel of water by a string attached to the base of the vessel: if the vessel be allowed to fall and be stopped suddenly when its velocity is v, find the impulsive tension of the string.

58. Prop. Body turning about a fixed point. *To find the conditions of equilibrium satisfied by a body wholly or partly immersed in a fluid and free to turn about a fixed point.*

Let O be the fixed point.

The forces acting on the solid are

(i) its weight W, vertically downwards through G, the centre of gravity;

(ii) the force of buoyancy W', vertically upwards through G', the centre of buoyancy;

(iii) the force at O.

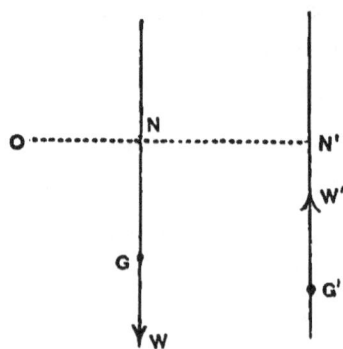

The three forces must be in the same plane, which must be a vertical one as (i) and (ii) are vertical.

FLOATING BODIES. 91

Hence O, G, G' must lie in the same vertical plane, (A).

Draw ONN' horizontal meeting the lines of action of (i) and (ii) in N, N' respectively.

Taking moments about O,
$$W \cdot ON - W' \cdot ON' = 0.$$

Hence G, G' must lie on the same side of the vertical through O and the horizontal distance of G from $O \times W$
$$= \text{that of } G' \times W' \quad\ldots\ldots\ldots\ldots\ldots\ldots\text{(B)}.$$

(A) and (B) constitute the necessary and sufficient conditions of equilibrium.

The action at O is obviously $W - W'$ upwards.

EXAMPLES.

1. A rectangle movable about an angular point floats with half its area immersed in a liquid. If the angular point lie outside the liquid, and if the rectangle float with its sides equally inclined to the vertical, shew that the ratio of the density of the rectangle to that of the liquid is $3b + a : 4b$ where a and b are the sides of the rectangle.
[Jesus Coll., 1883.]

2. A lamina in the form of a regular hexagon $ABCDEF$ can turn freely in a vertical plane about a hinge at A which is in the surface of a liquid. If in the position of equilibrium AB be above the fluid and half of BC be immersed, shew that the densities of the liquid and solid are in the ratio of 12 to 11.
[M. T., 1890.]

3. An equilateral triangle ABC, of weight W and specific gravity σ, is movable about a hinge at A, and is in equilibrium when the angle C is immersed in water and the side AB is horizontal. It is then turned about A in its own plane until the whole of the side BC is in the water and horizontal: prove that the action at the hinge in this position
$$= \frac{2(1 - \sqrt{\sigma})}{\sqrt{\sigma}} W. \qquad \text{[M. T., 1861.]}$$

*59. **Stability of Floating Bodies.** We shall now consider the conditions for the *stability* of a solid floating in liquid. If the solid be pushed further down in the liquid, the force of buoyancy is increased, and therefore tends to raise the body again; if the solid is raised, the force of buoyancy is diminished, and the weight will tend to sink the body again. For *vertical* displacements then, the equilibrium is *stable*.

It is assumed in the above that the solid is not floating in a fluid of the same density as itself: otherwise the equilibrium is *neutral*.

*60. Let us next consider displacements produced by an indefinitely small rotation about a horizontal axis.

We shall assume that the axis is such that the indefinitely small rotation about it does not alter the volume immersed, so that the force of buoyancy is unaltered in magnitude.

Let W be the weight of the floating body, G its centre of gravity. Let H be the centre of buoyancy, corresponding to the position of equilibrium, when HG is vertical. Let the plane of the paper be taken at right angles to the horizontal axis of rotation, and so as to contain HG. Let H' be the centre of buoyancy corresponding to the new position of the solid.

If H' be *not* in the plane of the paper, it is obvious that in the new position, the couple W at G, and W' at H' will produce rotation about a horizontal axis in the plane of the paper. We shall confine our attention entirely to the case in which H' is in this plane, when the couple will not cause rotation about a horizontal axis perpendicular to the first one.

It is obvious that if the plane of the paper be one of symmetry in the body, H' will lie in it. For instance, if a ship *pitch*, i.e. turn about a horizontal axis perpendicular to its keel, there will be no tendency to *rolling* produced thereby.

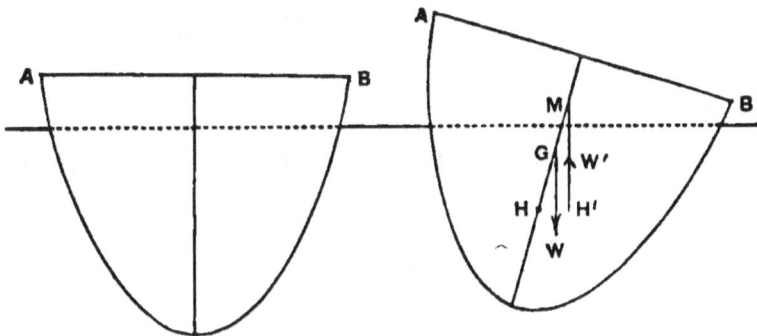

Let the vertical through H' when the body is in its displaced position meet HG in M.

Then, if M be *above* G, W upwards at H' and downwards at G, tend to *right* the body, i.e. to bring the body back to its old position. The equilibrium is therefore *stable*.

If M be below G, the reverse is the case, and the equilibrium is *unstable*.

If M coincides with G, the equilibrium is *neutral*.

The point M is called the **Metacentre** corresponding to the particular displacement made.

*61. DEF. **Surface of Buoyancy.** *If a body floating in a homogeneous liquid be supposed to take in turn every possible position for which the volume displaced remains constant, the locus of the centre of buoyancy is termed the Surface of Buoyancy.*

The section of the body made by the water-line in any position is termed the corresponding **Plane of Floatation.**

The surface enveloped by the planes of floatation is termed the **Surface of Floatation**.

*62. PROP. *The tangent-plane at any point of the surface of buoyancy is parallel to the corresponding plane of floatation.*

Let $AA'FBB'$ be the solid. Let H be the centre of buoyancy corresponding to the plane of floatation AOB.

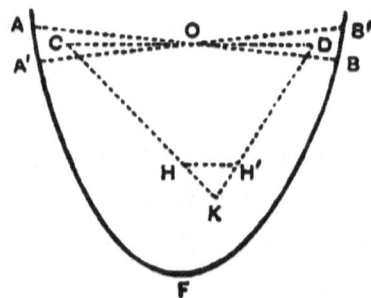

Let $A'OB'$ be a consecutive plane of floatation. Let H' be the corresponding centre of buoyancy.

Since the volume AFB = the volume $A'FB'$, that of the wedge AOA' = that of the wedge BOB'.

Let C, D, K be the centres of gravity of AOA', BOB', $A'FB$ respectively.

Join KC, KD. Then H, the centre of gravity of AFB must lie in CK, and divide it so that

$$CH : HK = \text{vol. } A'FB : \text{vol. } AOA'.$$

Similarly, H' is in KD, and

$$KH' : H'D = \text{vol. } BOB' : \text{vol. } A'FB$$
$$= HK : CH.$$

∴ HH' is parallel to CD, i.e. to AB, ultimately.

FLOATING BODIES. 95

In a similar way we can shew that the line joining any point on the surface of buoyancy near H with H is parallel to the plane AB; hence the tangent plane at H to the surface of buoyancy is parallel to the corresponding plane of floatation, AB.

63. PROP. *The positions of equilibrium of a floating solid are obtained by drawing normals from the centre of gravity of the solid to the surface of buoyancy.*

When the solid is in equilibrium the vertical through the corresponding centre of buoyancy passes through the centre of gravity. But this vertical line is perpendicular to the corresponding plane of floatation which is horizontal, and therefore by the last proposition is the normal to the surface of buoyancy. Hence the normal to the surface of buoyancy at every centre of buoyancy corresponding to a position of equilibrium passes through the centre of gravity of the solid.

It is obvious from the above that the positions of equilibrium of a floating solid correspond to those of a solid with the same centre of gravity, but bounded by the surface of buoyancy, and free to roll on a smooth horizontal plane.

We saw in Art. 60, that for certain rotations the lines through adjacent centres of buoyancy perpendicular to the corresponding planes of floatation meet, the point of intersection being the metacentre. A *metacentre* therefore is a point where two consecutive normals to the surface of buoyancy meet; it is in fact *a centre of curvature of the surface of buoyancy.*

64. Let us consider the case in which there is a plane of symmetry in the solid. This is the case with a cylinder, and a prism, also with a ship, since it is symmetrical about the vertical plane through its keel. If the displacements are confined to rotations about axes perpendicular to the plane of symmetry, the centres of buoyancy will lie on a curve in the plane of symmetry, which we may term the *Curve of Buoyancy*. From the last Art., the *locus of the metacentre* for these displacements will be the *evolute of the curve of buoyancy*.

EXAMPLES.

1. A uniform cylinder is floating with axis horizontal in a uniform liquid, find the curves of floatation and buoyancy, the number of positions of equilibrium, and the nature of the equilibrium, when (1) the cross section is a circle, (2) an ellipse, (3) a portion of a parabola cut off by a line perpendicular to the axis, the curved portion only being immersed.

Ans. The curves of floatation and buoyancy are (1) circles, (2) similar and similarly situated ellipses, (3) parabolas. The number of positions of equilibrium are (1) infinite, (2) 4, (3) 1.

2. If the immersed portion of a lamina floating with its plane vertical be (1) a triangle, (2) a rectangle, shew that the curve of buoyancy is (1) a hyperbola, (2) a parabola: find the curve of floatation.

3. A rectangular block of wood floats in mercury, the plane of floatation being a square, the side of which is 6 inches: the specific gravity of the mercury being 15 times that of the wood, shew that the equilibrium will be unstable if the height of the block exceeds 10 inches.
[M. T., 1854.]

4. A slender prism whose section is a square floats in a fluid with its axis horizontal, the ratio of its specific gravity to that of the fluid being as 3 to 4: find all the positions of equilibrium and determine which of them are stable and which unstable. [M. T., 1853.]

*65. PROP. *To determine the position of the metacentre corresponding to a rotation in a plane of symmetry in a floating body.*

Let the plane of the paper be the plane of symmetry.

FLOATING BODIES. 97

Let ACB be a plane of floatation, cutting off the volume V. Let H be the corresponding centre of

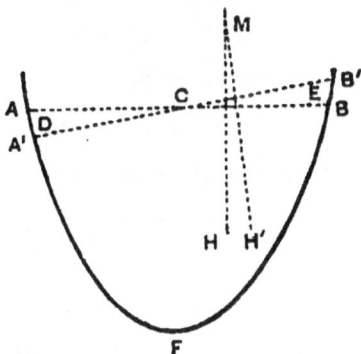

buoyancy. Let H' be the new centre of buoyancy when $A'CB'$ is the new plane of floatation, making an indefinitely small angle θ with ACB.

Since the vol. AFB = the vol. $A'FB'$,

the vol. ACA' = the vol. BCB'.

Draw HM perpendicular to ACB and $H'M$ perpendicular to $A'CB'$, then the limiting position of M is the metacentre.

Let D, E be the centres of mass of the volumes ACA', BCB' respectively.

The vol. $A'FB'$ = vol. AFB + vol. BCB' − vol. ACA', therefore taking moments about the line $H'M$,

$$V \cdot HM \sin\theta - \text{vol. } ACA' \cdot DE = 0 \quad \ldots\ldots(1),$$

$$\therefore HM = \mathrm{Lt}_{\theta=0} \frac{\text{vol. } ACA' \cdot DE}{V \sin\theta}.$$

This result may be expressed analytically as follows :—

Take the line ACB in the plane of the paper as axis of y, the perpendicular to the plane of the paper through

C that of x. Then if $dxdy$ be the area of an elementary portion of the area ACB, whose coordinates are x and y, the height of the small column on this portion, cut off by $A'CB'$ is $\pm y \tan \theta$, and the volume is $\pm y dx dy . \tan \theta$.

Hence instead of (1) we have
$$V . HM \sin \theta = \tan \theta . \iint y^2 dx dy,$$
where the integration is extended over the area ACB.
$$\therefore HM = \frac{\iint y^2 dx dy}{V} = \frac{AK^2}{V},$$
where AK^2 is the moment of inertia of the area ACB about the line through C perpendicular to the plane of the paper.

EXAMPLES.

1. Find the position of the metacentre in a right circular cylinder of radius a and length h, floating with axis vertical in a liquid whose density is $\frac{4}{5}$ that of the cylinder.

2. If the cross section of the cylinder in Ex. 1 be an ellipse of semi-axes a and b, find the metacentres corresponding to displacements in the planes of symmetry.

3. A uniform right circular cone of semi-vertical angle $30°$ is floating with its axis vertical and its vertex downwards in a liquid whose density is $\frac{2}{3}$ its own; determine whether the equilibrium is stable or unstable.

66. In Art. 60 we obtained the condition for the stability of a solid when the displacement is one of rotation through an indefinitely small angle. It is obvious however that a Naval Architect should be able to solve the problem of insuring the stability of a ship when the displacements are considerable.

Let ABC be the curve of buoyancy: let B be the centre of buoyancy corresponding to a position of equilibrium: let O be the corresponding metacentre, G the centre of mass of the solid. Let pOq be the evolute of ABC.

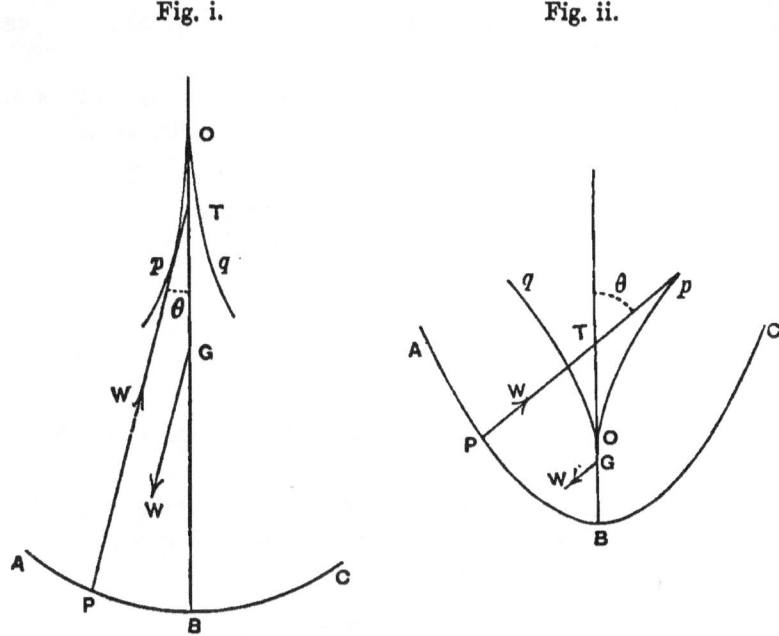

Fig. i. Fig. ii.

If the body be displaced through such an angle θ that PpT becomes vertical, the moment of the couple tending to *right* the body will be $W.GT\sin\theta$, and will be positive so long as T is above G, but not otherwise. It is obvious that in fig. i. T approaches G as the angle of displacement increases, whereas in fig. ii, T recedes from G as the angle increases. Hence the body is more likely to right itself for finite displacements in the case of a body whose curve of buoyancy is like that of figure ii, than in that of a body whose curve of buoyancy resembles that in fig. i.

ILLUSTRATIVE EXAMPLES.

1. *A prolate spheroid is totally immersed in a liquid with its axis at a given depth. Find the direction and magnitude of the resultant thrust upon one of the lower octants of its curved surface cut off by vertical and horizontal planes.* [Peterhouse, 1886.]

Let ABC be the octant, OA being the axis, OB a horizontal semi-diameter perpendicular to OA, and OC a vertical semi-diameter.

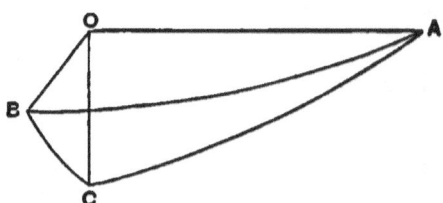

Let h be the depth of OA below the effective surface, a the length of OA and b that of OB and OC.

The vertical thrust on ABC is equal to the weight of the liquid which would fill the volume contained by the surface ABC and the vertical lines drawn through its perimeter to the surface.

The volume of this liquid $= \frac{\pi}{4} abh + \frac{\pi}{6} ab^2$.

The horizontal thrust in the direction OB on the octant = the thrust on OAC (Art. 53).

This equals the weight of the volume of the column of liquid which would stand on the area OAC and whose length is the depth of the centre of mass of OAC below the effective surface.

This volume $= \frac{\pi}{4} ab \times \left(h + \frac{4b}{3\pi} \right) = ab \left(\frac{\pi h}{4} + \frac{b}{3} \right)$.

Similarly the horizontal thrust in the direction OA on the octant = the weight of the volume $b^2 \left(\frac{\pi h}{4} + \frac{b}{3} \right)$.

As the components of the thrust in three directions at right angles are known, its magnitude and direction are known.

It should be observed that as the three forces to which the thrust on the octant is equivalent do not generally pass through the same point, they do not generally reduce to a single force.

2. *A regular solid tetrahedron, whose weight is equal to the weight of water it displaces, is completely immersed in water: shew that if it be cut into two halves by a central section parallel to two opposite edges, and one half be held fast, the force required to draw away the other half will always be the same, provided the centre of the tetrahedron is always in the same horizontal plane.* [M. T., 1878.]

Let $ABCD$ be the tetrahedron, $FGKH$ the central section parallel to the edges AD and BC.

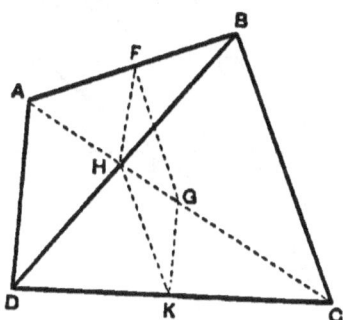

Let us consider the forces acting on $BCGH$, which prevent its being moved away from the other part. Its weight is the same as that of the liquid displaced: the thrust on it is the same as that on the liquid displaced except that on the latter there is a thrust across the face $HFGK$, and none in the case of the solid body. The liquid body can be moved away by an indefinitely small force, and therefore the solid body will require a force equal to the thrust across the face $HFGK$ on the liquid displaced. But this thrust depends on the area of the section $HFGK$ and on the depth of its centre of mass, which coinciding with the centre of the tetrahedron is at a constant depth.

Hence the required force is constant.

3. *An isosceles triangular lamina of density $n\sigma$ floats in two liquids of density σ and 2σ respectively, the depth of the upper liquid being b: prove that if the base of the lamina be not immersed, and be inclined at an angle θ to the horizon, θ must either be zero or be given by the equation*

$$(1+\cos 2\theta)\{n^3 a^6 (\cos 2\theta + \cos 2\alpha)^3 + 3n^2 a^4 b^2 (\cos 2\theta + \cos 2\alpha)^2 - 4b^6\}$$
$$= 2n^3 a^6 \cos^2 \alpha\, (\cos 2\theta + \cos 2\alpha)^4,$$

where 2α is the vertical angle, and a is a side of the lamina.

[M. T., 1886.]

102   FLOATING BODIES.

Let OAB be the triangle, O being its vertex at a depth $x - \tfrac{1}{2}b$ below the surface of the lower liquid, and at a depth $x + \tfrac{1}{2}b$ below that of the upper.

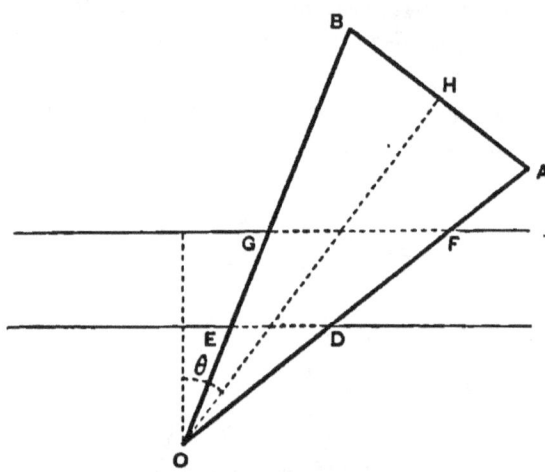

Let DE be the line in which the surface of the lower liquid cuts the triangle, FG that in which the surface of the upper liquid cuts it.

We may replace the force of buoyancy, which is the weight reversed of the liquid density 2σ displaced by ODE, and the liquid density σ displaced by $DEGF$, by the weight reversed of liquid density σ displaced by OFG, together with that of liquid density σ displaced by ODE.

The area ODE
$$= \tfrac{1}{2} OD \cdot OE \sin 2a = \tfrac{1}{2}(x - \tfrac{1}{2}b)^2 \sec(\theta - a) \sec(\theta + a) \sin 2a.$$
Similarly the area $OFG = \tfrac{1}{2}(x + \tfrac{1}{2}b)^2 \sec(\theta - a) \sec(\theta + a) \sin 2a.$

The area $AOB = \tfrac{1}{2} a^2 \sin 2a.$

The horizontal distance of c. m. of OED from O
$$= \tfrac{1}{3}\{OE \sin(\theta - a) + OD \sin(\theta + a)\} = \tfrac{1}{3}(x - \tfrac{1}{2}b)\{\tan(\theta - a) + \tan(\theta + a)\}.$$
Similarly that of c. m. of $OFG = \tfrac{1}{3}(x + \tfrac{1}{2}b)\{\tan(\theta - a) + \tan(\theta + a)\}.$

The horizontal distance of c. m. of OAB from $O = \tfrac{2}{3} a \sin\theta \cos a.$

Since the weight of body = force of buoyancy,
$$\tfrac{1}{2}\sigma\{(x - \tfrac{1}{2}b)^2 + (x + \tfrac{1}{2}b)^2\} \sec(\theta - a) \sec(\theta + a) \sin 2a = \tfrac{1}{2} n\sigma a^2 \sin 2a.$$

Since the centres of mass and buoyancy are in a vertical line
$$\tfrac{1}{2}\sigma\{(x - \tfrac{1}{2}b)^3 + (x + \tfrac{1}{2}b)^3\} \sec(\theta - a) \sec(\theta + a) \sin 2a \cdot \tfrac{1}{3}\{\tan(\theta - a)$$
$$+ \tan(\theta + a)\} = \tfrac{1}{2} n\sigma \cdot a^2 \sin 2a \cdot \tfrac{2}{3} a \sin\theta \cos a.$$

FLOATING BODIES. 103

These equations become

$$(x^2 + \tfrac{1}{4}b^2) = \tfrac{1}{4}na^2 (\cos 2a + \cos 2\theta) \dots\dots\dots\dots (1),$$

$$x^3 + \tfrac{3}{4}xb^2 = \tfrac{1}{4}na^3 (\cos 2a + \cos 2\theta)^2 \frac{\cos a}{\cos \theta} \dots\dots\dots (2).$$

By squaring (2) and substituting for x^2 from (1) we obtain the required relation.

4. *A cylindrical bucket with water in it balances a mass M over a pulley. A piece of cork, of mass m and specific gravity σ, is then tied to the bottom of the bucket so as to be totally immersed. Prove that the tension of this string will be*

$$\frac{2Mmg}{2M+m}\left(\frac{1}{\sigma}-1\right).\qquad \text{[M. T., 1873.]}$$

The mass on one side the pulley being $m+M$, and that on the other M, the former will descend with acceleration $g \cdot \dfrac{m}{m+2M}$.

The forces on the cork are

(i) its weight mg downwards,

(ii) the tension of the string T, downwards,

(iii) the force of buoyancy, upwards.

(iii) is equal to the resultant thrust on the water displaced, whose mass is m/σ.

But the weight of water displaced, mg/σ – resultant thrust would give the displaced water its resultant acceleration $\dfrac{mg}{m+2M}$;

$\therefore$ the resultant thrust $= \dfrac{mg}{\sigma} - \dfrac{m}{\sigma} \cdot \dfrac{mg}{m+2M}$.

The resultant force on the cork gives it its acceleration;

$$\therefore mg + T - \frac{mg}{\sigma} + \frac{m}{\sigma} \cdot \frac{mg}{m+2M} = m \cdot \frac{mg}{m+2M};$$

$$\therefore T = \frac{2Mmg}{2M+m}\left(\frac{1}{\sigma}-1\right).$$

5. *A solid cylinder floats in water in a cylindrical vessel, and the system revolves about the common axis with angular velocity ω. R and r being the radii of the vessel and the cylinder, shew that the cylinder is depressed by the motion through the space $\dfrac{\omega^2}{4g}(R^2 - r^2)$.*

[Smith's Prizes, 1842.

FLOATING BODIES.

It is assumed that the cylinder occupies a symmetrical position in the vessel. It is obvious that there is a symmetrical position of *actual* equilibrium; besides this, there may be positions of *relative* equilibrium, in which the cylinder rotates with the water.

Let bb' represent the plane surface of water, when there is no motion, BAB' the paraboloidal surface when the water is rotating.

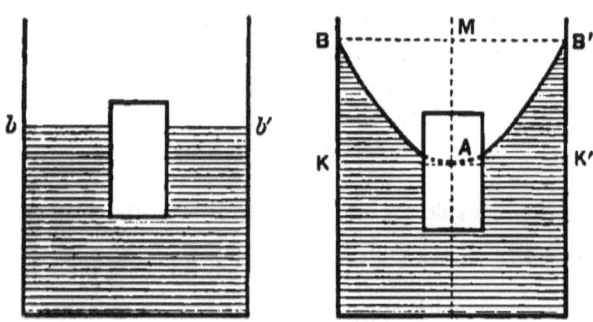

As the cylinder is in equilibrium in both cases, it displaces in each case its own weight of water, and the volume displaced is therefore constant.

Let the plane BB' meet the axis in M. Let the plane through A at right angles to the axis meet the vessel in KK'. Let x be the depth A has sunk below bb', and let y be the distance the cylinder has risen relatively to A.

Then $x - y =$ the actual distance through which the cylinder has sunk.

∵ the volume of water is constant,

$\pi R^2 \cdot x =$ vol. of water above $KK' = \frac{1}{2}$ vol. $BKK'B$

$$= \frac{1}{2} \pi R^2 \cdot AM = \frac{1}{2} \frac{\pi R^4 \omega^2}{2g},$$

$$\because BM^2 = \frac{2g}{\omega^2} AM;$$

$$\therefore x = \frac{R^2 \omega^2}{4g}.$$

Similarly, since the volume of water displaced by the cylinder is constant, it may be shewn that

$$y = \frac{r^2 \omega^2}{4g};$$

$$\therefore x - y = \frac{\omega^2}{4g}(R^2 - r^2).$$

FLOATING BODIES.

6. *Two equal uniform rods whose density is ρ are joined together at an angle 2α. If they be immersed with the angle downwards in a fluid of density σ, find the positions of equilibrium, and shew that the rods cannot rest with the line joining their extremities inclined to the horizon unless $(\sigma - \rho)/(\sigma + \rho)$ be greater than $\sin^2 \alpha$. If this condition be fulfilled, determine which of the positions of equilibrium is stable.* [M. T., 1861.]

Let AB, BC be the rods, each of length a. Then the total length of the two rods immersed in a position of equilibrium is $2b$, where $\sigma b = \rho a$. Let us find the corresponding curve of buoyancy.

Let BD be the line bisecting the angle ABC, and let PQ cut off a

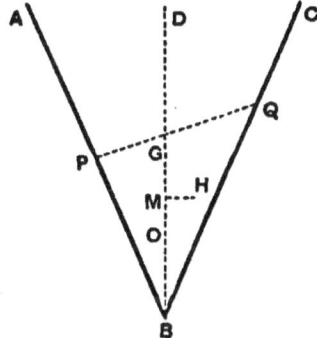

total length $2b$, so that $PB + BQ = 2b$. Let $PB = b - x$, $BQ = b + x$. Let H be the corresponding centre of buoyancy. Draw HM at right angles to BD.

Then $$HM = \frac{\frac{1}{2}(b+x)^2 \sin \alpha - \frac{1}{2}(b-x)^2 \sin \alpha}{2b} = x \sin \alpha,$$

and $$BM = \frac{\frac{1}{2}(b+x)^2 \cos \alpha + \frac{1}{2}(b-x)^2 \cos \alpha}{2b} = \frac{b^2 + x^2}{2b} \cos \alpha.$$

If $$OB = \tfrac{1}{2}b \cos \alpha, \quad OM = \frac{x^2}{2b} \cos \alpha.$$

$$\therefore HM^2 = 2b \cdot OM \sin \alpha \tan \alpha.$$

Hence the curve of buoyancy is a parabola whose vertex is O, axis OD, and latus rectum $2b \sin \alpha \tan \alpha$.

The centre of mass of the rods is at G, in BD, where $BG = \tfrac{1}{2} a \cos \alpha$. The positions of equilibrium are obtained by drawing normals from G to the curve of buoyancy.

It is obvious from the property of a parabola that if the distance of G from the vertex O is less than the semi-latus rectum, only one normal can be drawn from G to the parabola, i.e. the normal at O. If however the distance of G from O exceeds the semi-latus rectum, three can be drawn, one at O and two others, equally inclined to it on either side.

There will then be one or three positions of equilibrium

according as $\tfrac{1}{2}(a-b)\cos\alpha$ is $<$ or $> b\tan\alpha\sin\alpha$,

.... $(a-b)\cot^2\alpha$ is .... $2b$,

.... $(\sigma-\rho)\dfrac{1-\sin^2\alpha}{\sin^2\alpha}$ 2ρ,

.... $\dfrac{\sigma-\rho}{\sigma+\rho}$ .... $\sin^2\alpha$.

Since the locus of the metacentre is the evolute of the curve of buoyancy, in the symmetrical position of equilibrium the metacentre is above G, if OG is less than the semi-latus rectum, but not otherwise.

Hence when there are three positions of equilibrium, the symmetrical one is unstable, the other two stable; when there is only one position of equilibrium, it is stable.

The height of the metacentre above O might have been obtained by the method of Art. 65.

EXAMPLES. CHAPTER V.

1. An oblique cylinder standing on a horizontal plane, the generating lines making an angle α with the vertical, is filled to a height h with a weight W of liquid. Prove that the resultant thrust on the curved surface of the cylinder is equivalent to a couple of moment $\tfrac{1}{2}Wh\tan\alpha$ tending to upset the cylinder.

[M. T., 1884.]

2. An indefinitely thin hollow sphere A and a solid sphere B connected together are in equilibrium when totally immersed in a liquid: if A were solid and B hollow the combination would float in the same liquid with B half immersed. Prove that the radii of the spheres are in the ratio of $1 : \sqrt[6]{2}$. Also compare the density of the solid sphere with that of the liquid. [M. T., 1871.]

3. A double funnel formed by joining two equal hollow cones at their vertices stands upon a horizontal plane with the common axis vertical, and fluid is poured in until its surface bisects the axis of the upper cone. If the fluid be now on the point of escaping between the lower cone and the plane, prove that the weight of either cone is to that of the fluid it can hold as 27 : 16.

[M. T., 1861.]

4. A right circular cone of density σ whose base is an ellipse floats vertex downwards in a liquid of density ρ with the extremity of the shortest generator on the surface and its base inclined at an angle β to the surface. Prove that the longest generator is vertical, and that
$$1 + \tan a \tan \beta = (\rho/\sigma)^{\frac{2}{3}},$$
where a is the vertical angle of the cone. [M. T., 1879.]

5. A sphere is totally immersed in heavy fluid, and a line is drawn from the centre representing in magnitude and direction the resultant thrust on the surface of any hemisphere: shew that the locus of the extremity of this line is a sphere.

[M. T., 1876.]

6. A solid hemisphere of radius a and weight W is floating in liquid and at a point on the base at a distance c from the centre rests a weight w: shew that the tangent of the inclination of the axis of the hemisphere to the vertical for the corresponding position of equilibrium, assuming the base of the hemisphere entirely out of the liquid, is $\frac{8}{3} cw/aW$. [M. T., 1882.]

7. A hemispherical shell is floating on the surface of a liquid, and it is found that the greatest weight that can be attached to the rim is one-fourth of the weight of the hemisphere: prove that the weight of the liquid which would fill the hemisphere bears to the weight of the hemisphere the ratio of
$$25\sqrt{5} : 2\sqrt{50} - 28.$$
[M. T., 1880.]

8. A hollow copper spherical shell is floating just immersed in water at $0°C$. Prove that as the temperature rises the shell will again be just immersed at a temperature $8 + 3k/A$: k being the coefficient of expansion of copper, and the law of density of water being
$$\rho_t = \rho_4 \{1 - A(t-4)^2\}.$$

Prove also that the shell will be highest out of water at a temperature halfway between those for which it is just immersed.

[M. T., 1880.]

9. A solid cube made of uniform material can turn freely about one edge which is fixed in the surface of water: prove that if the cube rests with the face which is not immersed at an angle 30^0 to the horizon, the density of the cube is to that of the water in the ratio of $25 - 7\sqrt{3} : 18$. [M. T., 1889.]

10. A tetrahedron $DABC$ can float in a liquid of density greater than itself with either of the two edges DA, DB on the surface. Shew that either $DA = DB$, or that the tangents of the angles of the triangle ABC are in geometric progression. If the tetrahedron can also float with DC on the surface, it is regular. [M. T., 1879.]

11. A regular tetrahedron $ABCD$ is immersed with the face ABC vertical, the side AB being horizontal and in the surface of the liquid. CE is drawn perpendicular to AB meeting it in E. Shew that the line of action of the resultant thrust on the remaining faces of the tetrahedron divides CE in F, so that
$$EF : FC = 5 : 13.$$ [Pet., 1885.]

12. A right circular cone of height h and vertical angle $2a$, made of uniform material, floats in water with its axis vertical and vertex downwards and a length h' of axis immersed. The cone is bisected by a vertical plane through the axis and the two parts are hinged together at the vertex. Shew that the two halves will remain in contact if
$$\tan^2 a < h'/(h - h').$$ [M. T., 1886.]

13. A right circular cone has a plane base in the form of an ellipse: the cone floats on a fluid with its longest generator horizontal: if $2a$ be the vertical angle of the cone, and β the angle between the plane base and the shortest generating line, shew that
$$\cot \beta = \cot 4a - \tfrac{1}{5} \operatorname{cosec} 4a.$$ [M. T., 1866.]

14. In H.M.S. *Achilles*, a ship of 9000 tons displacement, it was found that moving 20 tons from one side of the deck to the other, a distance of 42 feet, caused the bob of a pendulum 20 feet long to move through 10 inches. Prove that the metacentric height was 2·24 feet. [M. T., 1884.]

15. Prove that, if a solid ellipsoid float in fluid of twice its density, it must float with a principal plane as the plane of floatation, and that, if it float with the largest axis downwards, the equilibrium is entirely unstable: if with the mean axis downwards, stable for one principal displacement and unstable for the other; and if with the smallest axis downwards, entirely stable.

[M. T., 1882.]

16. A basin, formed of the segment of a spherical surface, is movable about a horizontal axis, which is a diameter of the base of the segment. Prove that the basin will upset if the ratio of the weight of water poured in to the weight of the basin is greater than the ratio of d to $D - 2d$, where d is the depth of the basin, and D the diameter of the sphere from which it is cut. [M. T., 1883.]

17. A cone is suspended by its vertex from a point above the surface of a liquid, and rests with a generating line vertical. If the vertical angle of the cone be 60° and the height of the vertex above the liquid equal to the radius of the base, prove that the densities of the cone and liquid are in the radius of

$$2\sqrt{2} - 1 : 2\sqrt{2}.$$ [M. T., 1868.]

18. A homogeneous body, which can move round a fixed horizontal hinge, is at rest partly immersed in a homogeneous fluid. If the level of the fluid can be altered until the same plane section of the body can remain at rest in the surface of the fluid, prove that the density of the fluid must be twice that of the body.

[M. T., 1864.]

19. A uniform lamina in the form of an equilateral triangle floats with its plane vertical, shew that there will be only one position of equilibrium with a given one of the angles immersed and the opposite side entirely out of the fluid unless the ratio of the density of the fluid to that of the lamina lies between 16/9 and 2.

[Pet., 1890.]

20. Two closely fitting hemispheres made of sheet metal of small uniform thickness are hinged together at a point on their rims, and are suspended from the hinge, their rims being greased so that they form a water-tight spherical shell: this shell is now filled with water through a small aperture near the hinge: prove

that the contact will not give way if the weight of the shell exceed three times the weight of the water it contains. [M. T., 1887.]

21. A sphere of density σ floats just immersed in three liquids. The densities of the liquids in descending order are ρ, 4ρ and 9ρ, and the thicknesses of the two upper liquid layers are each one-third of the sphere: prove that $27\sigma = 122\rho$. [Clare Coll., 1890.]

22. A wooden sphere of radius r is held just immersed in a cylindrical vessel of radius R containing water, and is allowed to rise gently out of the water: prove that the loss of potential energy of the water is
$$Wr(3R^2 - 2r^2) \div 3R^2,$$
W being the weight of water displaced by the sphere.

[St John's Coll., 1881.]

23. Prove that a sphere partly immersed in a basin of water cannot rest in stable equilibrium on the summit of any convex portion of the base. [Pet., 1889.]

24. A bucket half-full of water is suspended by a string which passes over a pulley small enough to let the other end fall into the bucket. To this end is tied a ball whose specific gravity σ is greater than 2. Shew that, if the ball do not touch the bottom of the bucket and if no water overflow, equilibrium is possible if the weight of the ball lie between W and $\sigma W/(\sigma - 2)$, where W is the weight of the bucket and water. [Pet., 1888.]

25. Two spherical shells of the same material and of thicknesses proportional to their radii are each half-filled with water. Shew that when tied to the ends of a string slung over a smooth pulley, and allowed to fall, the resultant thrusts of the water on the spheres are equal. [M. T., 1890.]

26. A homogeneous solid floats in liquid: if, when the temperature of both is raised by the same amount, the depth of the lowest point of the solid remain unaltered, then the coefficients of cubical expansion of the solid and liquid are in the ratio of $3M$ to $3M - M'$, where M is the mass of the solid, and M' the mass of a cylindrical volume of liquid of base equal to the area of the plane of floatation and height the constant depth of the lowest point. [M. T., 1891.]

FLOATING BODIES. 111

27. A cylindrical piece of wood of length l and sectional area a is floating with its axis vertical in a cylindrical vessel of sectional area A which contains water: prove that the work which is done in very slowly pressing down the wood until it is just completely immersed is
$$\tfrac{1}{2}gal^2(1-a/A)(\rho-\sigma)^2/\rho,$$
where ρ and σ denote the densities of the water and wood respectively. [M. T., 1889.]

28. A regular tetrahedron has one edge fixed in the surface of a fluid. Shew that it will be in equilibrium with the other edge inclined to the vertical at an angle $\operatorname{cosec}^{-1} 3$, if the density of the tetrahedron is to that of the fluid as 19 : 64. [M. T., 1881.]

29. A triangular lamina of known weight floats in water so that its plane is not vertical, with its centre of gravity in the surface and one angular point immersed, this point being in contact with a rough fixed surface: find the specific gravity of the lamina and the reaction at the angular point. [M. T., 1876.]

30. Prove that a solid body of any law of density bounded by a spherical and a plane surface cannot have more than one position of equilibrium with the plane fully immersed unless it have an infinite number: and that in the case of a homogeneous hemisphere of density ρ and radius a in fluid of density σ the position is stable and the metacentric height
$$\tfrac{3}{8}a(\sigma-\rho)/\rho. \qquad [\text{M. T., 1888.}]$$

31. A square lamina (side $2a$) has one angular point in a fluid and rests in a vertical plane on two smooth horizontal pegs in the surface of the fluid and at a distance c apart. If the specific gravity of the material of the lamina compared with water is $c^2/24a^2$, prove that the inclination of a side of the square to the horizon in an unsymmetrical position of equilibrium is given by
$$\cos 2\theta \cos(\theta+\tfrac{1}{4}\pi)=a/c\sqrt{2}. \qquad [\text{Pet., 1886.}]$$

32. The frustum of a right circular cone bounded by planes perpendicular to the axis is totally immersed in water with its axis inclined at an angle ψ to the vertical, and ϕ is the inclination to the vertical of the direction of the resultant thrust on the curved surface of the frustum: θ is the corresponding inclination

when the axis of the frustum is horizontal, with the vertex of the cone at the same depth as before: prove that

$$\tan \phi = \frac{3 \sin 2\psi - 2 \tan \theta \sin \psi}{1 + 3 \cos 2\psi - 2 \tan \theta \cos \psi}.$$ [M. T., 1886.]

33. A cone floats in liquid which fills a fixed conical shell: both the cone and the shell have their axes vertical and vertices downwards: the vertical angles of the cone and shell are equal and the axis of the shell is twice that of the cone. If the cone be pressed down until its vertex very nearly reaches the vertex of the shell, so that some of the liquid overflows, and then released, it is found that the cone rises until it is just wholly out of the liquid and then begins to fall. Prove that the densities of the cone and the liquid are in the ratio $45 - 21\sqrt[3]{7} : 4\sqrt[3]{7}$, the free surface of the liquid being supposed to remain horizontal throughout the motion.

[M. T., 1875.]

34. A vessel of thin material in the form of a paraboloid of revolution contains liquid and floats in another liquid, shew that the equilibrium will be always stable provided the liquid inside be denser than that without, the mass of the vessel being supposed small. [Peterhouse, 1885.]

35. A solid right cone of density σ, height h, and vertical angle $2a$ can turn freely about its vertex which is fixed at a height d above the surface of a liquid of density ρ. If it float with its base wholly immersed, and its axis inclined obliquely at an angle θ with the vertical, shew that

$$h^4 (\rho - \sigma) \{\cos(\theta + a) \cos(\theta - a)\}^{\frac{3}{2}} = d^4 \rho \cos \theta \cos^3 a.$$

[Pet., 1890.]

36. A solid homogeneous cone with an elliptic base floats in a homogeneous fluid with its longest generator horizontal and immersed in the fluid and the centre of the base in the surface. If a is the length of the horizontal generator and h the height of the highest point above it, prove that the line of action of the resultant thrust on the curved surface cuts the vertical through the centre of gravity of the cone at a depth

$$(3\pi + 16)(4a^2 + 25h^2)/800h$$

below the highest point; find the magnitude of this thrust.

[M. T., 1884.]

FLOATING BODIES.

37. A portion of a homogeneous elliptic cylinder, the eccentricity of a right section of which is $\frac{1}{2}$, is bounded by one of the planes through the latera recta of the right section, and floats in homogeneous liquid with its axis in the surface, and no part of the bounding plane immersed. Shew (i) that the density of the liquid is to that of the cylinder as

$$8\pi + 3\sqrt{3} : 12\pi,$$

(ii) that there are three positions of equilibrium, of which two are stable. [M. T., 1884.]

38. A right circular cone of semi-vertical angle a floats, vertex downwards, in water contained in a vertical cylinder, and the surface of the water meets the cone in a circle of radius r. The water is made to rotate with uniform angular velocity ω about the common axis of the cone and cylinder: shew that the water now meets the cone in a circle of radius r', given by

$$r'^3 - \frac{3\omega^2}{4g} r'^4 \tan a = r^3. \qquad \text{[Trin. Coll., 1890.]}$$

39. A fine parabolic tube, whose weight may be neglected, has its plane vertical, so that it is free to roll on a horizontal table: if the tube contain some liquid, prove that there is only one position of equilibrium, and that unstable. [M. T., 1865.]

40. A mass of liquid not acted on by gravity revolves uniformly round a fixed axis, and contains, revolving with it, two small solids connected by a string, one of the solids being denser and the other rarer than the liquid: find the condition of equilibrium and discuss the cases which may arise.

[Smith's Prize, 1865.]

41. Two liquids which do not mix are placed in a vertical cylinder of radius c, so that the lighter liquid (density ρ') lies above the heavier liquid (density ρ) and occupies a portion of the cylinder of length k. Into this cylinder is gently lowered a solid (density σ) in the shape of a portion of a paraboloid of revolution of latus rectum $2c$ cut off by a plane perpendicular to the axis and passing through the focus. If the paraboloid just floats in the liquid vertex downwards when completely immersed, prove that

$$(\rho - \sigma) c = 4 (\sqrt{ck} - k)(\rho - \rho'). \qquad \text{[Pet., 1891.]}$$

G. E. H.

42. A vessel in the form of a cube, of side $12a$ containing liquid is placed so as to rest on the top of a perfectly rough sphere of radius $5a$: neglecting the weight of the vessel, prove that for displacements in planes parallel to the vertical faces, there will be stability provided the depth of the liquid is between $4a$ and $6a$.

[M. T., 1881.]

43. The displacement of a laden ship is W tons, and a is the height of the metacentre above its centre of gravity. Some deck cargo of weight w is moved across the deck and the ship tilts through an angle θ. This cargo is now removed to another deck distance h below the upper one and placed below its second position on the upper deck. Prove that the ship now tilts through an angle

$$aW\theta/(aW+hw).$$ [Trin. Coll., 1890.]

44. Liquid of density ρ is standing in a fixed smooth circular cylinder with axis vertical of a radius a. This is made to revolve about the axis with uniform angular velocity ω, none of the base being exposed. A paraboloidal solid of density σ, shaped just to fit the cavity in the liquid, is gently placed upon the surface so that its flat top just passes through the highest rim of the liquid. If $\rho > \sigma$, shew that before it reaches its equilibrium position, the liquid rising round it, it must sink through a depth

$$\{1-(\rho-\sigma)^{\frac{1}{3}}/\rho^{\frac{1}{3}}\}^2 \omega^2 a^2/4g,$$

supposing no interference with the base to take place.

[M. T., 1885.]

45. A rectangular parallelepiped (edges $2a$, $2b$, $2c$) of density τ, floats with its edges $2a$ vertical, partly in a lower liquid of density ρ and partly in an upper liquid of density σ. Prove that for displacements in the planes parallel to $2a$, $2b$, the equilibrium is stable, if

$$b^2(\rho-\sigma)^2 > 6a^2(\rho-\tau)(\tau-\sigma),$$

ρ, τ, σ being in descending order of magnitude, and the parallelepiped being wholly immersed. [Peterhouse, 1888.]

46. A bridge of boats supports a plane rigid roadway AB in a horizontal position. When a small movable load is placed at G the bridge is depressed uniformly: when the load is placed at a point C the end A is unaltered in level; when at D the end B

is unaltered in level: and when at P the point Q of the roadway is unaltered in level. Prove that

$$AG \cdot GC = BG \cdot GD = PG \cdot GQ:$$

and that the deflection produced at a point R by a load at P is equal to the deflection produced at P by the same load at R.

[M. T., 1883.]

47. A sphere of density $m\sigma$ is immersed in a large vessel of water of density σ so that a length c of its vertical diameter is under water. The centre is then depressed through a depth h. Find an expression for the gain in potential energy of the sphere and water together. Hence prove that, if the sphere were originally floating at rest, the gain in potential energy (h being small) would be to a first approximation

$$\tfrac{1}{2}\pi\sigma g h^2 a^2 (1-z^2),$$

where z is the middle root of the equation

$$(z+1)^2(z-2)+4m=0. \qquad \text{[M. T., 1886.]}$$

CHAPTER VI.

THE DETERMINATION OF SPECIFIC GRAVITY.

67. IN order to ascertain the specific gravity of a given substance we must find the weights of equal volumes of the substance and the standard substance. The weight of a body can be measured with very great accuracy, but it is often impracticable to measure its volume *directly*. In this case indirect methods must be adopted. One method is to employ the **Specific Gravity Bottle.**

The *Specific Gravity Bottle* is a glass bottle with the inside of its neck and its glass stopper carefully ground so that when pushed home the stopper always occupies the same position. A perforation runs through the stopper so that, when the bottle is quite full of liquid and the stopper is pushed in, the excess of liquid escapes through the perforation and allows the stopper to go quite home.

68. PROP. *To determine the specific gravity of a liquid by means of the specific gravity bottle.*

Observations to be made:

(i) Find the weight of the bottle when empty; let this be W.

(ii) Find the weight of the bottle when full of water; let this be W':

∴ the weight of the water filling the bottle is $W' - W$.

THE DETERMINATION OF SPECIFIC GRAVITY. 117

(iii) Find the weight of the bottle when full of the given liquid; let this be W''':

∴ the weight of the given liquid filling the bottle is $W''' - W$.

Hence the specific gravity of the liquid

$$= \frac{\text{weight of the liquid filling the bottle}}{\text{weight of water filling the bottle}} = \frac{W''' - W}{W' - W}.$$

69. Prop. *To determine the specific gravity of a solid by means of the specific gravity bottle.*

Observations to be made—

(i) Find the weight of the solid, W.

(ii) Find the weight of the bottle when full of water; let this be W'.

(iii) Find the weight of the bottle, when it contains the solid and is filled up with water; let this be W''':

∴ $W''' - W'$ = the weight of the solid − the weight of water it displaces:

∴ the weight of water displaced by the solid

$$= W - W''' + W'.$$

Hence the specific gravity of the solid $= W/(W + W' - W''')$.

It is obvious that this method will not apply unless it is convenient to break up the solid into portions small enough to go into the bottle.

70. Since as a rule the experiments of Arts. 68 and 69 are not made *in vacuo*, it may be desirable to make corrections on account of the buoyancy of the air. Thus in Art. 68, if the experiments are made in air, $W' - W$ is really the excess of the weight of water that will fill the bottle over the weight of the air: as the specific gravity of air is known and the internal volume of the bottle is known approximately from the weight of water, we can estimate the actual weight of water it will hold. For air being comparatively very light, a small error in estimating the internal volume of the bottle will not affect the weight of the air materially. Similarly we can correct for the weight of the liquid.

Also in Art. 69, W is not the actual weight of the solid but is the excess of that weight over the weight of air the solid displaces. As we have already ascertained the approximate specific gravity of the solid, we can calculate the weight of air it displaces, and so deduce the actual weight of the solid.

EXAMPLES.

1. A specific gravity bottle full of water weighs 50·3 grammes. Some pieces of metal weighing 17·6 grammes are introduced and the bottle is again filled with water, the combined weight being 65·9 grammes. Find the specific gravity of the metal.

2. A specific gravity bottle full of oil weighs 42·5 grammes. A lump of metal weight 11·2 grammes is placed in the bottle and the bottle filled up with oil, the whole now weighing 52·1 grammes. The bottle is now emptied and 20 grammes of mercury (sp. gr. 13·5) poured into it: on filling up with oil, the combined weight is found to be 61·9; find the specific gravities of the oil and the metal.

71. The U-tube Method. By this method we can compare the specific gravities of two liquids, which when placed in contact, have, like oil and water, or mercury and water, a clearly defined surface of separation.

Let the two liquids be poured, one down each of the two branches of a U-shaped tube of glass.

THE DETERMINATION OF SPECIFIC GRAVITY. 119

Let A be the free surface of one liquid, of specific gravity ρ, and A' that of the other of specific gravity ρ'. Let B be the surface of separation.

Let h, h' the vertical heights of A and A' above B be observed, either by means of a graduated rule, or better by means of a cathetometer, an instrument for measuring the vertical distance between two points.

Since AB is a homogeneous liquid,

the pressure at $B =$ atmospheric pressure $+ g\rho h$, and since $A'B$ is a homogeneous liquid,

the pressure at $B =$ atmospheric pressure $+ g\rho' h'$;

$$\therefore g\rho h = g\rho' h',$$
$$\therefore \rho/\rho' = h'/h;$$

or the specific gravities are inversely proportional to the heights of the free surfaces above the surface of separation.

72. The inverted U-tube method. By this method we can compare the specific gravities of any two liquids.

Take a piece of glass tubing of the shape indicated in the figure, and fix it so that the two lower open ends dip one into each of the two vessels A and B containing the liquids. Attach a piece of india-rubber tubing to the upper opening C, and suck the liquids some distance up the two branches. Observe the height of the liquid surface in each branch above that in the corresponding vessel. It is obvious that these heights are inversely proportional to the specific gravities.

This apparatus is practically *Hare's hydrometer*.

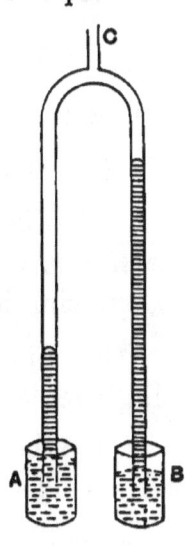

120 THE DETERMINATION OF SPECIFIC GRAVITY.

73. The Hydrostatic Balance. This consists of an ordinary Balance, one scale-pan of which is suspended by shorter chains than the other: the first scale-pan has a hook attached to its under surface so that a solid can be suspended from it and immersed in a vessel containing some liquid.

74. PROP. *To determine the specific gravity of a solid by means of the Hydrostatic Balance.*

(α) *When the solid is heavier than water.*

Observations to be made—

(i) Weigh the solid; let W be its weight.

(ii) Weigh the solid, when it is immersed in water; let W' be this weight:

∴ the weight of water displaced by the solid is $W - W'$ (Art. 58);

∴ the specific gravity of the solid is $W/(W - W')$.

(β) *When the solid is lighter than water.*

Observations to be made—

(i) Find the weight W of the solid.

(ii) Find x the weight in water of a piece of metal, the *sinker*, which when attached to the solid will cause both to sink in water.

(iii) Find y the weight in water of the sinker and solid together;

∴ y = the weight of the sinker in water + the weight of the solid − the weight of water displaced by the solid

$= x + W -$ the weight of water displaced by the solid;

THE DETERMINATION OF SPECIFIC GRAVITY. 121

∴ the weight of water displaced by the solid
$$= W + x - y;$$
∴ the specific gravity of the solid $= W/(x + W - y)$.

75. Prop. *To determine the specific gravity of a liquid by the Hydrostatic Balance.*

Observations to be made—

(i) Find the weight W of a solid which will sink both in the liquid and in water.

(ii) Find the weight W' of the solid in water.

(iii) Find the weight W'' of the solid in the given liquid.

∴ $W - W' =$ the weight of water displaced by the solid, and $W - W'' = $ .... given liquid ....
∴ the specific gravity of the liquid $= (W - W'')/(W - W')$.

76. In the above cases it is supposed that the solid is weighed *in vacuo*: if however W is its weight in air, we ought strictly to make a correction for the weight of air it displaces. Knowing the specific gravity of air and using the uncorrected formula we can find approximately the specific gravity of the solid, and hence the weight of air it displaces. The true weight of the solid can then be deduced.

If the solid is one that will melt in water, its specific gravity may be compared with that of some liquid in which it does not melt. Or, it may be coated with wax and the specific gravity of the whole found: then, as the weight and specific gravity of the wax can be found, we can deduce that of the solid. The last method also applies to a solid like pumice-stone which absorbs water.

122 THE DETERMINATION OF SPECIFIC GRAVITY.

77. *Jolly's Balance.* This is practically a hydrostatic balance in which a spring balance is used instead of a common one. It consists of a long spiral spring to which is attached a pan. From this pan is hung by a fine wire a second pan which is kept immersed in water. The body whose specific gravity is required is first placed in the upper pan and the compression produced in the spring observed by means of a scale. The body is removed and the same compression produced by weights placed in the upper pan. This gives the weight in air. The weight is next placed in the lower pan and weights placed in the upper until the compression is the same as before. This gives the weight of water displaced by the body. We can obviously apply the method of Art. 75 to obtain the specific gravity of a liquid by this balance.

EXAMPLES.

1. The apparent weight of a piece of platinum in water is 60 grammes and the absolute weight of another piece of platinum twice as big as the former is 126 grammes. Determine the specific gravity of platinum.

2. A piece of wood weighing 5 oz. has attached to it a piece of copper weighing 4 oz. and of specific gravity 9: the weight of the two together in turpentine of specific gravity ·9 is found to be 2·5 oz.: what is the specific gravity of the wood?

3. A piece of silver and a piece of gold are suspended from the two arms of a balance which is in equilibrium when the silver is immersed in alcohol (density = ·85) and the gold in nitric acid (density = 1·5). The densities of the silver and gold being 10·5 and 19·3 respectively, what are their relative masses?

4. A uniform cylinder floats in mercury with 5·14 inches of its axis immersed. Water is poured on the mercury to the depth of 1 inch and it is then found that 5·066 inches of the axis are below the surface of the mercury. Find the specific gravity of mercury.

5. If W, W' be the weights of a body in vacuo and in water respectively, shew that its weight in air of specific gravity s will be

$$W - s(W - W'). \qquad \text{[Jesus Coll., 1886.]}$$

6. The apparent weight of a sinker, weighed in water, is four times the weight in vacuum of a piece of a material, whose specific gravity is required: that of the sinker and the piece together is three times that weight. Shew that the specific gravity of the material is ·5.

[M. T., 1855.]

THE DETERMINATION OF SPECIFIC GRAVITY. 123

78. The Common Hydrometer. This instrument is used to determine the specific gravity of liquids. It consists of a uniform thin straight stem ending in a bulb A, cylindrical or spherical: below A is another small bulb B weighted, generally with mercury, so that the hydrometer will float with the stem vertical. This hydrometer is generally made of glass. When it is floating in a liquid it displaces its own weight of the liquid, so that the lighter the liquid, the greater the volume submerged. This fact enables us to graduate the stem so that we read off the specific gravity from the graduation in the surface of the liquid. As the weight of the liquid displaced is always the same, this is sometimes called the *constant weight* hydrometer.

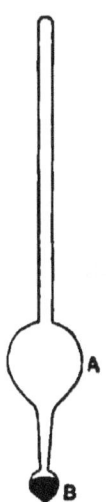

79. *To graduate the common hydrometer.*

Let W be the weight, V the volume of the hydrometer, α the sectional area of the stem: let x, x' be the lengths of the stem unimmersed, when the hydrometer floats in liquids of specific gravities s, s' respectively. Then since the volume of liquid displaced is $V - \alpha x$,

$$s(V - \alpha x) = W = s'(V - \alpha x') = V - \alpha a \text{ (Art. 55)},$$

where a is the unimmersed portion of the stem when floating in water, the standard substance.

$$\therefore a = V/\alpha - W/\alpha,$$

and
$$x = V/\alpha - W/\alpha s;$$

$$\therefore x - a = \frac{W}{\alpha}\left(1 - \frac{1}{s}\right) \quad \dots\dots\dots\dots\dots\dots(\text{i}).$$

Similarly, $$x' - a = \frac{W}{\alpha}\left(1 - \frac{1}{s'}\right);$$

$$\therefore \frac{x' - a}{x - a} = \frac{1 - 1/s'}{1 - 1/s} \quad \text{......(ii).}$$

Let the hydrometer float in water and mark the position of the water-line on the stem: next let it float in any convenient liquid of known specific gravity s, and mark the new line of floatation; the distance between the two marked points is $x - a$. Hence the distance $x' - a$, between the first mark and the graduation for specific gravity s' is obtained by equation (ii).

By giving s' different values we can obtain any number of graduations.

80. It can also be shewn that *the graduations corresponding to specific gravities in Arithmetic Progression are at distances from a certain point in the stem produced, which are in Harmonic Progression.*

Let AB be the stem of the hydrometer, C the graduation corresponding to water (sp. gr. 1), P that corresponding to a liquid of sp. gr. s.

$$\underline{A \quad\quad C \quad P \quad\quad\quad\quad\quad B \quad\quad\quad\quad O}$$

Then using the notation of the last article

$$AP = x \text{ and } AC = a.$$

∴ equation (1) becomes

$$CP = W/a \cdot (1 - 1/s) \quad \text{......(iii).}$$

Produce AB to a point O, such that $CO = W/a$, i.e. CO is a length of the stem, which would displace a weight W of water.

∴ (iii) becomes
$$CP = CO\,(1 - 1/s).$$

$$\therefore OP = \frac{OC}{s}.$$

EXAMPLES.

1. If a common hydrometer floats in water, with 3·3 inches of its stem unimmersed, and in liquid of specific gravity ·85 with 1·2 inches unimmersed, find the specific gravity of a liquid in which it floats with 6 inches unimmersed, and also the specific gravity of a liquid in which it floats with 2 inches unimmersed.

2. Assuming the stem of a common hydrometer to be of uniform section, and that the highest graduation corresponds to a specific gravity of 1·00 and the lowest to 2·00, find what specific gravity corresponds to the position of the point half-way between the two divisions. Find also the graduation corresponding to 1·50.

81. Nicholson's Hydrometer. This can be used for ascertaining the specific gravity of either a liquid or a solid. It can also be used for weighing small bodies. It consists of a hollow cylinder A, generally made of brass, with a thin wire stem supporting a cup B above it. Another cup C below A is connected with it by a wire: C is weighted so that the hydrometer may float with the stem vertical. There is a well-defined mark D on the stem and the instrument is always weighted so that when floating in any liquid, D is in the surface. Since the volume immersed is always the same, this hydrometer is sometimes called the *constant volume* hydrometer.

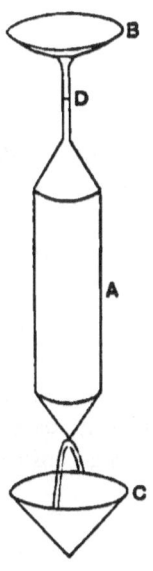

82. *To compare the specific gravity of one liquid with that of another.*

Find (i) W, the weight of the hydrometer;

(ii) w, the weight that must be placed in B so

that the hydrometer may float in liquid of specific gravity s with D in the surface;

(iii) w', the weight that must be placed in B so that the hydrometer may float in liquid of specific gravity s' with D in the surface.

∴ if V be the volume immersed in either case
$$Vs = W + w, \text{ and } Vs' = W + w';$$
$$\therefore s'/s = (W + w')/(W + w).$$

83. *To find the specific gravity of a solid.*

Place the hydrometer in water and put weights in B so that D is in the surface. Next put the solid in B and take out weights W so that D is again in the surface;

∴ W is the weight of the solid.

Now take the solid out of B and put it in C, and add weights W' until D is again in the surface.

W' must be the weight of the water displaced by the solid;

∴ the specific gravity of the solid $= W/W'$.

<small>Owing to capillarity, the water meets the hydrometer stem either as at A, or as at B, and in order that our results may be accurate, we must ensure that the way is the same in all our observations.</small>

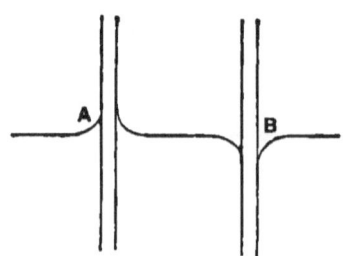

THE DETERMINATION OF SPECIFIC GRAVITY. 127

EXAMPLES.

1. If the weight required to sink a Nicholson's hydrometer is 500 grains and when an amalgam of gold, specific gravity 19·4, and of mercury, specific gravity 13·56, is placed in the two cups 264 and 278 grains respectively are required: find the volumes of the two metals. [Clare Coll., 1885.]

2. It is found that a weight w of a solid, placed in the upper cup, sinks a Nicholson's hydrometer to the same depth in water as a weight W of the same substance sinks it when placed in the lower cup: supposing w, W are the actual weights in vacuo, while the other experiments have been performed in the presence of the atmosphere of specific gravity s, shew that the specific gravity of the solid is

$$(W - ws)/(W - w).$$

EXAMPLES. CHAPTER VI.

1. Having determined by the hydrostatic balance method the specific gravity of a thin wire, find the greatest error in determining the mean transverse section of a wire ten yards long, if the weights are determined accurately to tenths of a grain and the weight of a cubic inch of water is 252·5 grains. [M. T., 1870.]

2. A mixture of $100 + x$ parts by weight of alcohol and $100 - x$ of water is said to be x above proof: find the specific gravity of such a mixture in terms of that of alcohol, and shew how to graduate the common hydrometer in such a manner that its reading for any mixture may give the excess of the mixture above proof.

[M. T., 1876.]

3. A common hydrometer being graduated upwards, its readings for two different fluids are x_1 and x_2, and for a mixture of equal parts of these x: shew that the volume of a unit of length of the stem is to the volume of the whole instrument below the zero point as

$$x_1 + x_2 - 2x \; : \; xx_1 + xx_2 - 2x_1 x_2. \quad \text{[M. T., 1872.]}$$

4. Shew that if σ be the specific gravity of a substance found by the use of the hydrostatic balance when the specific gravity of the air is neglected, and s be the specific gravity of air, then the error will be $s(\sigma - 1)$.

5. A Nicholson's hydrometer is used to determine the weight and specific gravity of a solid, and W and σ are the results when the effect of the air is neglected: prove that the actual weight is

$$W\{1+a/\sigma\,(1-a)\}\,(1-a/\rho),$$

where a and ρ are the specific gravities of air and of the material of the known weights employed.

6. If the reading of a common hydrometer when placed in fluid at the same temperature as itself be x, and if, when it is placed in the same fluid at a higher temperature than itself, its reading be at first x_1, but afterwards the reading rises to x_2, the ratio of the expansions of the fluid and of the hydrometer for the same change of temperature is approximately

$$x-x_1\ :\ x_2-x_1. \qquad \text{[M. T., 1882.]}$$

7. A common hydrometer is used to determine the specific gravity of a liquid which is at a temperature higher than that of water. When the hydrometer is transferred from water to the liquid the specific gravity appears at first to be σ but afterwards to be σ_1. Shew that the true specific gravity at the temperature of the water is

$$\sigma+\frac{a'}{a}\,(\sigma_1-\sigma),$$

where a and a' are the coefficients of expansion of the hydrometer and the fluid respectively. [Trin. Coll., 1891.]

8. A common hydrometer has a piece of its bulb chipped off. When placed in fluids of density a and β, it indicates densities a', β' respectively. Find the proportion of the weight which has been chipped off, and shew that if in any fluid the apparent density is x, the true density is

$$\frac{xa\beta\,(\beta'-a')}{a'\beta'\,(\beta-a)-x\,(a'\beta-a\beta')}. \qquad \text{[M. T., 1890.]}$$

9. Supposing a common hydrometer to be immersed in a liquid less dense than water as far as the point to which it would sink in water, prove that, if let go, it will sink through a distance

$$\frac{2\omega}{k}\cdot\frac{1-s}{s},$$

ω being the weight of the hydrometer, k the section of its stem, and s the specific gravity of the liquid: the hydrometer is supposed to be never entirely immersed. [M. T., 1880.]

10. An old Nicholson's hydrometer is found with its stem uniformly coated with rust. Two weights of unknown magnitude are also found with it. The stem has three marks A, B, C upon it which marked the surface of some unknown liquid when the hydrometer (free from rust) floated in it either free, or with one or other of the two weights in its upper pan. When it is now placed in a liquid, the surface in the three cases is at A', B', C'. Shew that
$$AA' \cdot BC - AC \cdot BB' + AB \cdot CC' = 0. \quad \text{[M. T., 1878.]}$$

11. If s be the value obtained for the specific gravity of a liquid by means of the specific gravity bottle and a spring balance, the air displaced being neglected, and σ the value obtained when the air displaced is taken into account, if water be taken to be 815 times as heavy as air, prove that
$$\sigma = s - \frac{s-1}{815}. \quad \text{[Peterhouse, 1890.]}$$

12. The weight of a specific gravity bottle is found when full of glycerine to be x at $t°$ C. and x' when at $t'°$ C.: when a piece of lead weighing w is placed inside the bottle, and the bottle filled up with glycerine the total weight is y when at $t°$ C., and y' when at $t'°$ C. Assuming that the bottle, the glycerine and the lead expand uniformly when raised in temperature, find the relative rates of expansion of glycerine and lead. [Jesus Coll., 1893.]

CHAPTER VII.

ON GASES.

84. WE have already seen (Arts. 4 and 6) what property gases and liquids have in common as fluids, and also what their distinguishing characteristics are.

On weighing a flask full of air, and then exhausting the air, and weighing again, Otto von Guericke found that the empty flask weighed somewhat less than when full. This shewed that *the air has weight*.

85. That *the air exerts pressure* is verified by pushing a tumbler mouth downwards into a vessel of water; it will be found that the surface of the water inside the tumbler is below that outside.

If the air exerts pressure, it must follow that if the air is not allowed access to one portion of the surface of a liquid, while it is in contact with another portion, its pressure will force the latter down below the level of the former. This is the principle of the *Barometer*, which affords the most satisfactory way of measuring the pressure of the air.

The Barometer.

86. *To construct a Barometer.*

Take a uniform straight glass tube, closed at one end and about 32 inches long. Fill this with mercury, close the open end with the thumb and invert the tube, placing the temporarily closed end below the surface of some mercury in a trough: remove the thumb and it will be found that the mercury will remain in the tube, so that, at the sea-level, its surface is between 29 and 31 inches above that in the trough.

Let B be the surface of the mercury in the tube; let A be a point in the surface of the mercury outside.

If σ be the density of the mercury, the pressure at A exceeds that at B by $g\sigma h$, where h is the vertical height of B above A. (Art. 21.)

But the pressure at A is the pressure of the air.

Hence the pressure of the air exceeds the pressure in the tube above B by $g\sigma h$. Now if by careful heating all air and moisture have been expelled from the mercury in the tube, and none allowed to get in when the tube was inverted, there can be nothing to exert pressure in the space above B except the vapour of mercury.

The pressure exerted by mercury vapour has been shewn by Regnault's and Haager's experiments to be very small, so that we may take the pressure of the air to be $g\sigma h$. The space in the tube above B is termed the *Torricellian Vacuum*.

87. *Corrections to be applied to an observation with the mercury barometer.*

(1) *Correction for capacity.*

As the mercury rises in the tube, it will fall in the trough, and vice-versâ. The distance generally measured in a barometric observation, is the height of the mercury surface in the tube above some selected standard position. Let x be this observed distance, and let c be the height of the mercury surface in the tube above that in the trough, when the former is in the standard position. Then when the one rises through a distance x the other must sink through a distance xa/A, where a, A, are the sectional areas of the tube and trough respectively. Hence the required height of the surface in the tube above that in the trough is

$$c+x+xa/A,$$

i.e. $+xa/A$ is the correction.

(2) *Correction for variation in the density of the mercury on account of temperature.*

Since mercury expands with a rise of temperature, the barometer will rise as the temperature rises, without any change in the atmospheric pressure having taken place.

In order that the height of the barometer may be proportional to the atmospheric pressure, the density of the mercury should be constant. We must, therefore, calculate the height of the column which the atmospheric pressure would support if the mercury were at some standard temperature, 0° C. for instance. If h' be this required height, h the observed height, and σ_0, σ_t, the densities of mercury at 0° C. and t° C., the actual temperature at the time of observation,

$$\sigma_0 h' = \sigma_t h,$$

also
$$\sigma_0 = \sigma_t (1+\lambda t),$$

where λ is the coefficient of expansion of mercury.

$\therefore\ h' = h/(1+\lambda t) = h(1-\lambda t)$, approximately;

$\therefore$ the correction is $-h\lambda t$.

(3) *Correction for expansion or contraction of the measuring rod.*

Since the measuring rod used to determine the apparent height x of the mercury surface in the tube above the standard position expands or contracts as the temperature rises or falls, it will only give this height correctly at the standard temperature.

If the temperature at the time of observation be t° C. above the standard temperature, and k be the coefficient of expansion of the rod, what appears to be x is really $x+xkt$,

$\therefore$ the correction is $+xkt$.

ON GASES. 133

(4) *Correction for capillarity.*

It is a well-known experimental fact that when a fine tube open at both ends is pushed down into mercury the mercury surface is lower in the tube than outside. This depression is due to capillarity, and depends on the material of the tube and the bore, varying inversely as the diameter of the bore. On this account the height of the mercury in the barometer tube is less than it would be if there were no capillarity, or if the sections of the tube and trough were the same. The correction to be added is a *constant* positive one for each barometer, and may be obtained experimentally, or calculated from the bore of the tube if the sort of glass of which the tube is made is known.

(5) *Correction for height above sea-level.*

If the observation is taken above the sea-level, the height is less than it would have been if it had been taken at the same time and place but at the sea-level.

By Art. 98, this correction is $h(e^{\frac{gz}{k}} - 1)$, where z is the altitude above sea-level, and k is the constant of Art. 91.

88. Other liquids besides mercury may be used in the construction of a barometer; the advantages attaching to mercury are due to

(i) its great density: this renders a comparatively short tube only necessary. If water were employed, the height of the column would be over 32 feet.

(ii) its small alteration in volume for a change of temperature.

(iii) the very small pressure exerted by its vapour. The pressure exerted by water-vapour is considerable, and varies considerably with the temperature.

89. The average height of the barometer is about 76 centimetres, or 29·922 inches. The corresponding atmospheric pressure is about 14·7 lbs. per square inch, and is termed an *Atmosphere*. Large pressures are generally measured by engineers in atmospheres; it should, however, be remembered that this system of measurement is not scientific, as the value of an atmosphere varies from place to place, for the same reason that a lb.-weight varies.

90. *The Aneroid Barometer.* This instrument contains a chamber from which the air has been exhausted, and one side of which consists of a metallic diaphragm. The other side of the diaphragm is exposed to the air, and any variation in the air-pressure causes a deflection in the diaphragm, which is communicated to an index on a dial. The dial is graduated by comparing it with a mercury barometer.

The advantages of this barometer are its small size and its convenient shape. The disadvantage is that it is impossible to ensure the same accuracy as in a mercury barometer.

91. The following law due to Robert Boyle, by whose name it is known in England, gives the relation between the pressure and density of a gas, when the temperature is constant.

Boyle's Experimental Law.

The pressure of a gas varies directly as its density, or inversely as the volume occupied, if the temperature remain constant.

The law may be conveniently verified as follows:

Take a straight glass tube AB of uniform bore, and closed at the end A. Connect this tube with another similar tube CD, open at both ends, by a piece of flexible india-rubber tubing. Now partly fill the tube with mercury, and then closing the open end C with the thumb, incline the tubes so that the air bubbles up to the end A.

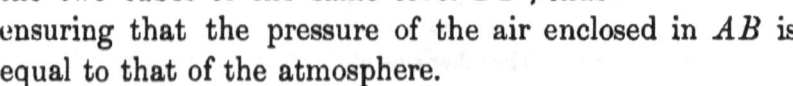

Next clamp both tubes in a vertical position, as in the figure, the ends A and C being uppermost, and remove the thumb from C. By raising or lowering the tube CD, we can bring the mercury surfaces in the two tubes to the same level PP', thus ensuring that the pressure of the air enclosed in AB is equal to that of the atmosphere.

ON GASES. 135

Observe (i) the height of the barometer, h,

(ii) the length AP of the tube, occupied by the air; this length is proportional to the volume.

Now raise CD so that the mercury-surfaces in AB, CD are Q, Q' respectively: Q' will clearly be above Q.

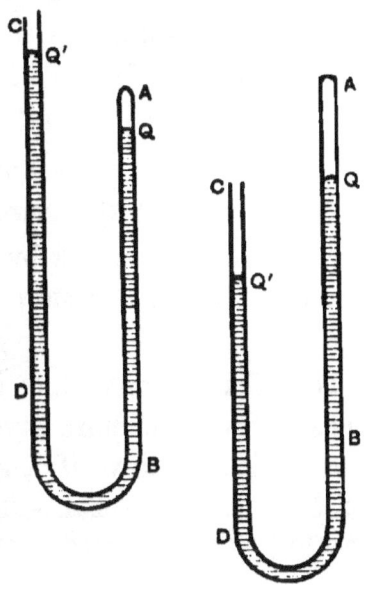

Observe (iii) h' the difference of level between Q and Q';

(iv) the length AQ.

Make similar observations when CD is lowered so that Q' is below Q.

If a considerable number of such observations be taken, it will be found that the product

$$AQ(h+h') \text{ or } AQ(h-h'),$$

according as Q is below or above Q', is always approximately equal to $AP \cdot h$.

But AQ is proportional to the volume of the air, and $h + h'$ or $h - h'$, as the case may be, to the pressure.

Hence the pressure × volume is constant, *i.e.* the pressure varies inversely as the volume.

But since the mass of air imprisoned remains constant, the density varies inversely as the volume, and therefore directly as the pressure.

Hence the law may be written

$$p = k\rho,$$

where ρ is the density, and k is a constant depending on the nature of the gas and the temperature.

The value of k may be ascertained by determining the mass and pressure of a known volume of the gas at the given temperature.

It would not be difficult to introduce into AB, instead of air, any other gas, hydrogen or pure oxygen, for instance, for which it might be desired to verify the law.

The apparatus described above is a modification of that used by Boyle; it is not so easy to describe as the original one, but much simpler to work with.

92. A *Perfect Gas* is defined as one which obeys Boyle's law absolutely. Air, oxygen, hydrogen and nitrogen may be regarded as very approximately perfect gases, since they obey the law very nearly for a considerable range of pressures and temperatures. A vapour in contact with its own liquid is so far from obeying the law that its pressure is quite independent of the volume. A diminution of volume is accompanied by a condensation of some of the vapour and an expansion by an evaporation of some of the liquid, without any alteration in pressure. If the expansion go on after all the liquid is evaporated, the pressure begins to diminish but not so rapidly as if it followed Boyle's law. But the larger the volume becomes the more nearly does the vapour approximate to a perfect gas.

When a gas is spoken of, it is generally assumed, unless stated otherwise, that it is a perfect gas.

The Diving-Bell.

93. This consists of a large bell-shaped or cylindrical vessel made of metal, capable of holding several persons. It is closed at the top and open underneath. It can be lowered into water by means of a chain: when this is done the water rises inside the bell and compresses the air, which will however always prevent the water from completely filling the bell.

It is usually provided with two tubes, through one of which fresh air can be pumped to any extent into the bell, driving if need be all the water out of the bell. The other is to draw off the vitiated air when necessary.

It is obvious that the lower the bell sinks, the greater is the pressure inside, and, if no fresh air is pumped in, the smaller the volume of water displaced, and in consequence the greater the tension of the chain supporting it.

The chief object of a diving-bell is to enable work such as the laying of the foundations of piers, etc. to be done under water.

EXAMPLES.

1. A conical wineglass is immersed mouth downwards in water: how far must it be depressed in order that the water within the glass may rise half-way up it? [M. T., '59.]

2. At a depth of 10 feet in a pond the volume of an air-bubble is ·0001 cu. inches: find approximately what it will be when it reaches the surface, if the height of the barometer is 30 inches, and the specific gravity of mercury 13·5. [Peterhouse, '88.]

3. A diving-bell is suspended at a fixed depth: a man who has been seated in the bell suddenly falls into the water and floats. Determine the effect on (1) the tension of the chain, (2) the level of the water in the bell, (3) the amount of water in the bell. [Jesus Coll., '88.]

4. A heavy sphere is placed in a vertical cylinder, filled with atmospheric air, which it exactly fits. Find the density of the air in the cylinder when the sphere is in a position of permanent rest.

[M. T., '57.]

5. A cylindrical diving-bell is lowered to a given depth in water by means of a chain and is completely immersed. If it be lowered to the same depth in a liquid of less specific gravity, will the tension of the chain be increased or diminished? [Peterhouse, '90.]

6. A cylindrical diving-bell whose volume is 450000 c. cms. is lowered in water to a depth of 1500 cms. and it is found that an addition of 700000 c. cms. of air at atmospheric pressure is required to fill the bell. Find the height of the water barometer and the pressure on the surface of the water in dynes per sq. cm., the value of g being taken as 980.

[Trinity Coll., '85.]

7. A barometer stands at 30 inches, and the space occupied by the Torricellian vacuum is 2 inches: if now a bubble of air which would at atmospheric pressure occupy half an inch of the tube be introduced into the tube, prove that the surface of the mercury in the tube will be lowered 3 inches. Shew also that the height of a correct barometer when this false one stands at x inches is $x + \dfrac{15}{32-x}$. [Jesus Coll., '85.]

8. A diving-bell is made of a substance whose specific gravity is 4, and its interior will contain a quantity of water whose weight is twice that of the bell: if the bell be lowered in water till the tension of the rope is half the weight of the bell, prove that the density of the air within it will be eight times that of the atmosphere. [M. T., '70.]

9. A diving-bell is in the form of a cylinder with a hemispherical top, c is the length and a the radius of the cylinder. Find how far the bell must be sunk in order that the hemisphere may be the only part containing air: shew that in that position the volume of air at atmospheric pressure which must be forced in, to clear the whole bell from water, must be $(c/H + \frac{2}{3}c/a) \times$ the volume of the bell, H being the height of the water barometer. [Clare Coll., '84.]

Charles' Law.

94. The following law, generally attributed to Charles, gives the connection between the volume and temperature of a gas when the pressure remains constant.

If the pressure of a gas remain constant, the volume increases by 1/273 of the volume at 0° C. for every degree Centigrade the temperature is raised.

The gases for which this law holds very approximately are the so-called *permanent gases*, oxygen, hydrogen, and nitrogen. The law does not apply to vapours.

If V_t, V_0 denote the volume of a gas at $t°$ C. and $0°$ C. respectively, the pressure being the same in each case, it follows from Charles' Law that

$$V_t = V_0 + \frac{V_0 t}{273},$$

$$\therefore V_t = V_0 \frac{273 + t}{273}.$$

If now the temperature be measured from a point 273° below the zero of the centigrade scale, we obtain what is termed the **absolute temperature**, the new zero being termed the **absolute zero.** If T be the absolute temperature corresponding to the temperature t on the ordinary scale

$$T = t + 273,$$

$$\therefore V_t = \frac{V_0 T}{273}.$$

Hence we may state the law thus

The volume of a gas, when the pressure is constant, varies as the absolute temperature.

95. From the laws of Boyle and Charles we can deduce the relation between the pressure and temperature of a gas, when the volume is constant.

Let P, V, and T denote the pressure, volume and absolute temperature of a mass of gas.

By Boyle's Law
$$V \propto \frac{1}{P}, \text{ when } T \text{ is constant.}$$
By Charles' Law
$$V \propto T \text{ when } P \text{ is constant.}$$
$$\therefore V \propto \frac{T}{P} \text{ when } P \text{ and } T \text{ both vary,}$$
$$\therefore VP \propto T, \dots\dots\dots\dots\dots\dots\dots\dots\dots\dots\dots\dots(i)$$
and $\quad P \propto T$ when V is constant, $\dots\dots\dots(ii)$.

Hence from (ii)

The pressure of a gas varies as the absolute temperature, when the volume is constant.

Since the volume of a given mass of gas varies inversely as its density, we may write instead of equation (i)
$$P = k\rho T,$$
where ρ is the density of the gas and k is a constant depending on the gas.

For experimental methods of verifying these laws the student is referred to the chapter on the *Dilatation of gases* in Prof. Balfour Stewart's *Treatise on Heat*.

Mixture of Gases.

96. Experimental Fact. *If two gases, occupying different vessels, be at the same temperature and pressure, they will, when one vessel is allowed to communicate with the other, form a mixture whose pressure is the same as before, provided no chemical action takes place.*

PROP. *If the pressures of two gases at the same temperature and volume be p, p' respectively, the pressure of the mixture at the same temperature and volume will be*
$$p + p'.$$

ON GASES. 141

Let v be the volume of each gas.

Let the pressure of the second gas be altered from p' to p, then by Boyle's Law its volume becomes vp'/p.

Now let the two gases at pressure p communicate with one another, the total volume will be $v + vp'/p$ or $v(p+p')/p$ and the pressure p.

If the mixture be compressed to volume v the pressure by Boyle's law will become $p+p'$.

EXAMPLES.

1. A volume of air of any magnitude, free from the action of force, and of variable temperature, is at rest: if the temperatures at a series of points be in arithmetical progression, prove that the densities at these points are in harmonical progression. [M. T., '57.]

2. Two vessels contain air having the same pressure Π but different temperatures t, t': the temperature of each being increased by the same quantity, find which has its pressure most increased. If the vessels be of the same size, and the air in one be forced into the other, find the pressure of the mixture at a temperature zero. [M. T., '56.]

3. A gas saturated with vapour is at a pressure Π. It is then compressed without change of temperature to $1/n$th its former volume, and the pressure is then observed to be $=\Pi_n$. Shew that the pressure of the vapour $=(n\Pi - \Pi_n)/(n-1)$, and that the pressure of the air in the original volume without its vapour $=(\Pi_n - \Pi)/(n-1)$. [Clare Coll., '88.]

4. The volume, pressure and temperature of a given quantity of gas are respectively v_1, p_1, t_1 and those of a second quantity of gas are v_2, p_2, t_2: supposing the two quantities to be mixed in a vessel of volume v, and then brought to a temperature t, find the pressure of the mixture. [S. John's Coll., '87.]

5. The weight of a litre of dry air at $0°$ C. and 760 mm. pressure is 1·293187 grammes, find the weight in grammes of V litres of dry air at $t°$ C. and p mm. pressure. [Peterhouse, '90.]

6. Find in ft.-lb. sec. units the value of k in the formula
$$p = k\rho(1+at),$$

for a gas of which at temperature $t°$ C. the mass of V cubic feet is W lbs., the height of the barometer being h inches, and the specific gravity of mercury compared to water being σ. [M. T., '91.]

7. Masses m, m' of two gases in which the ratios of the pressure to the density are respectively κ and κ' are mixed at the same temperature. Prove that the ratio of the pressure to the density in the compound is
$$(m\kappa + m'\kappa')/(m+m').$$
[M. T., '93.]

8. If a cylindrical diving-bell of height a and whose chamber could contain a weight W of water be lowered so that the depth of the highest point is d, prove that when the temperature is raised $t°$, the tension of the supporting chain is diminished by
$$\frac{1}{1+at} \cdot \frac{What}{\sqrt{(h+d)^2 + 4ah}} \text{ nearly,}$$
h being the height of the water barometer, and a the expansion of air for 1° C. [M. T., '69.]

9. Shew that if $p_1, \rho_1, t_1, p_2, \rho_2, t_2$ and p_3, ρ_3, t_3 be three corresponding pressures, densities, and temperatures of a perfect gas,
$$t_1(p_2/\rho_2 - p_3/\rho_3) + t_2(p_3/\rho_3 - p_1/\rho_1) + t_3(p_1/\rho_1 - p_2/\rho_2) = 0.$$
[M. T., '54.]

10. A cylindrical diving-bell of internal volume v is filled with air at atmospheric pressure Π and absolute temperature t, and is lowered to a certain depth below the surface of some water. Shew that if a very small rise (x) in the temperature, and increase (y) in the atmospheric pressure now take place, the apparent weight of the bell will be unaltered, provided $x\Pi v = tyv'$, v' being the volume of air in the bell.

[Jesus Coll., '80.]

Density of the Atmosphere.

97. Prop. *To determine the relation between the density of the air and the altitude.*

Let z denote the difference of altitude between two stations.

Let the air between the two stations be divided into n strata of equal thickness, z/n: let the densities at the lower sides of these strata, beginning at the lowest, be $\rho_1, \rho_2, \ldots \rho_n$; let ρ be the density at the upper face of

the highest stratum. If the temperature be assumed constant throughout, the corresponding pressures will be
$$k\rho_1, k\rho_2, \ldots k\rho_n, k\rho \quad \text{(Art. 91.)}$$

If n be taken indefinitely great, we may suppose the density of each stratum constant throughout, and the same as at the lower side.

Hence by Art. 21.
$$k\rho_1 - k\rho_2 = g\rho_1 \cdot z/n, \quad k\rho_2 - k\rho_3 = g\rho_2 \cdot z/n \ldots,$$
$$k\rho_{n-1} - k\rho_n = g\rho_{n-1} \cdot z/n, \quad k\rho_n - k\rho = g\rho_n \cdot z/n.$$
$$\therefore \rho_2 = \rho_1\left(1 - \frac{g}{k} \cdot \frac{z}{n}\right), \quad \rho_3 = \rho_2\left(1 - \frac{g}{k} \cdot \frac{z}{n}\right) \ldots,$$
$$\rho_n = \rho_{n-1}\left(1 - \frac{g}{k} \cdot \frac{z}{n}\right), \quad \rho = \rho_n\left(1 - \frac{g}{k} \cdot \frac{z}{n}\right).$$

Hence *as the altitude increases in Arithmetic the density diminishes in Geometric progression.*

Also equating the product of the left-hand expressions to that of the right-hand expressions, we have
$$\frac{\rho}{\rho_1} = \left(1 - \frac{g}{k} \cdot \frac{z}{n}\right)^n = e^{-\frac{g}{k} \cdot z},$$
since n is taken indefinitely great.

By using the Calculus, the proof can be simplified thus
$$\frac{dp}{dz} = -g\rho \quad \text{(Art. 26)},$$
and since $p = k\rho$,
$$\frac{1}{\rho}\frac{d\rho}{dz} = -\frac{g}{k};$$
$$\therefore \log \rho = -\frac{gz}{k} + \text{constant};$$
$$\therefore \rho = Ce^{-\frac{gz}{k}},$$
where C is a constant.

When $\quad z = 0, \quad \rho = \rho_1;$

$\quad\quad\quad \therefore \rho_1 = C,$

and $\quad\quad \dfrac{\rho}{\rho_1} = e^{-\frac{gz}{k}}.$

If this formula represented correctly the law of density, it would follow that the density of the atmosphere would not be zero except at an infinite distance from the Earth's surface. As, however, it is obtained on the supposition that the temperature and g are both constant throughout, neither of which suppositions is even approximately true for large differences of altitude, we cannot regard the formula as approximately correct except for small variations in altitude.

It is obvious that there must be a finite limit to the height of the atmosphere from the consideration that at a certain finite height the force of gravity is insufficient to prevent a particle from flying off on account of the Earth's rotation.

98. Prop. *To determine the height of one station above another by means of a barometric observation at each.*

Let h, h_1 be the heights of the barometer at the upper and lower stations respectively; z the required difference of altitude.

Let p, p_1 be the pressures, ρ, ρ_1 the densities at the two stations.

Then if the temperature be constant throughout

$$\dfrac{h}{h_1} = \dfrac{p}{p_1} = \dfrac{\rho}{\rho_1} = e^{-\frac{g}{k}z} \text{ (Art. 97)}.$$

$$\therefore z = \dfrac{k}{g} \log \dfrac{h_1}{h}.$$

99. The pressure at any point of the air is the weight of the column of air of gradually dwindling density, which stands on a unit area placed horizontally at the point. If we suppose this column to be compressed so that its density is throughout the same as at the lowest point, its new height is termed the *height of the homogeneous atmosphere* at the point.

Let ρ be the density of the air at the point in question, p the pressure there, and H the height of the homogeneous atmosphere.

Then
$$g\rho H = p,$$
$$\therefore H = p/g\rho = k/g = \text{constant},$$

if the temperature is constant.

Hence *the height of the homogeneous atmosphere is the same at all places at the same temperature.*

Giving k and g their average values at the sea-level, we ascertain that the height of the homogeneous atmosphere is a little less than 5 miles.

The result of Art. 97, may be written

$$\frac{\rho}{\rho_1} = e^{-\frac{z}{H}}.$$

ILLUSTRATIVE EXAMPLES.

1. *The readings of a perfect mercurial barometer are a and β, while the corresponding readings of a faulty one, in which there is some air, are a and b: prove that the correction to be applied to any reading c of the faulty barometer is*

$$\frac{(a-a)(\beta-b)(a-b)}{(a-c)(a-a)-(b-c)(\beta-b)}.$$

[M. T., '76.]

ON GASES.

Let k be the length of the tube of the false barometer above the mercury and occupied by air, when the true barometer stands at a. The pressure of the confined air in this position is measured by $a - a$, the difference between the true and faulty barometers.

When the true barometer stands at β, and the faulty one at b, the length of tube occupied becomes $k + a - b$, and the pressure is measured by $\beta - b$. Let $c + x$ be the height of the true barometer when the other stands at c; then the length of tube occupied by air is $k + a - c$, and the pressure is measured by x. Since the volume of the air is proportional to the length of the tube it occupies, and the product of the pressure and volume is constant (Boyle's Law),

$$(a - a)\, k = (\beta - b)(k + a - b) = x\,(k + a - c);$$

eliminating k, these give us the required value x, the correction to be added to c.

2. *Prove that, neglecting variations in temperature and gravity, the difference of gravitational potential energy of two equal masses of air is equal to the work necessary to compress the air at constant temperature to the volume of the other.* [M. T., '88.]

Let p, v be the initial pressure and volume of an indefinitely small portion of one mass, ρ its density; let this small portion be lowered a small distance z vertically, so that in its new position the pressure, volume and density are p', v' and ρ' respectively.

The mass of the element is $v\rho$, and therefore the change in its gravitational potential energy $= g\rho v z$.

The work done in compressing the element is $(v - v')\, p$.

But $vp = v'p'$, and also $p' - p = g\rho z$ (Art. 21).

$\therefore$ the work done $= v'\,(p' - p) = g\rho v' z$

$\qquad\qquad\qquad =$ change in gravitational energy,

since v' and v are ultimately equal.

Since the theorem holds for an indefinitely small mass when it is lowered an indefinitely small distance, we can by summing up the results for an infinitely large number of such masses and such distances, extend it to a finite mass lowered a finite distance.

ON GASES. 147

EXAMPLES. CHAPTER VII.

1. The readings of a faulty barometer containing some air are 29·4 and 29·9 inches, the corresponding readings of a correct instrument being 29·8 and 30·4 inches respectively: prove that the length of the tube occupied by the air is 2·9 inches, when the reading of the faulty barometer is 29 inches: and find the corresponding correct reading. [M. T., 1879.]

2. If a body floats on a liquid with volumes V_1, V_2, V_3 above the surface when the barometric heights are respectively h_1, h_2, h_3, prove that

$$h_1 V_1 (V_2 - V_3) + h_2 V_2 (V_3 - V_1) + h_3 V_3 (V_1 - V_2) = 0.$$

[Clare Coll., 1889.]

3. A thin heavy cylinder, hollow, and open at its lower end, is found when depressed from the atmosphere successively into three liquids, to remain at rest when its higher end is at respective depths h, h', h'' below the surfaces. If s, s', s'' be the specific gravities of the fluids, prove that, the weight of the air contained in the cylinder being neglected in comparison with that of the cylinder,

$$s(s' - s'')h + s'(s'' - s)h' + s''(s - s')h'' = 0.$$

[M. T., 1864.]

4. Air is compressed in a vessel at a pressure p and at the same temperature as the atmosphere. An aperture is then opened and shut the instant the air inside is at atmospheric pressure P, and it is found that when the air left in the vessel is again at the same temperature as the atmosphere its pressure is p'. Find how much air has issued and the temperature at the instant the aperture was shut. [M. T., 1875.]

5. A constant pressure air thermometer having been graduated when the barometer stood at 29 inches, find the true absolute temperature when the barometer is at 30 inches, the absolute temperature indicated by the instrument being 290°. [M. T., 1874.]

10—2

6. Determine the numerical value of pv/T in Fahrenheit's scale for a pound of air, supposing that a cubic foot of air is ·0763 lb. at a temperature of 62° F., when the barometer is 30 inches high: taking an inch as the unit of length, the weight of a pound as the unit of force, the density of mercury relative to water as 13·6, and − 273° C. as the absolute zero of temperature. [M. T., 1883.]

7. From the following data obtain the true reading of a barometer placed at a height 20 feet above sea-level:

Barometer reading 29·5 inches. Attached thermometer 20° C. Ratio of area of section of tube to section of cistern = 1 : 41. When the mercury in the cistern is at the zero of the scale, supposed marked on the tube, the mercury in the tube stands at 30 inches. Capillary action + ·04 in. Fall in barometric height for each foot above sea-level ·001 in. Coefficient of expansion of mercury for 1° C. = ·00018. [M. T., 1892.]

8. Two bulbs containing air are connected by a horizontal glass tube of uniform bore, and a bubble of liquid in this tube separates the air into two equal quantities. The bubble is then displaced by heating the bulbs to temperatures $t°$ and $t'°$: prove that if the temperature of each bulb be increased τ degrees, the bubble will receive an additional displacement which bears to the original displacement the ratio

$$2a\tau \;:\; 2 + a\,(t + t' - 2\tau),$$

where a is the coefficient of expansion for air. [M. T., 1868.]

9. Prove that the mass of the atmosphere is approximately equal to that of an ocean of mercury covering the earth, and of depth equal to the mean height of the barometer, and that this mass is $5·3 \times 10^{15}$ tons, the density of mercury being 13·6.

[M. T., 1884.]

10. A cylindrical diving-bell of height 10 feet and internal radius 3 feet is immersed in water so that the depth of the top is 100 feet. Shew that if the temperature of the air in the bell be now lowered from 20° C. to 15° C., and if 30 feet be the height of the water barometer at the time, then the tension of the chain is increased by about 67 lbs. [M. T., 1890.]

11. A box is filled with a heavy gas at uniform temperature. Prove that, if a is the altitude of the highest point above the lowest, and p, p' are the pressures at these points, the ratio of the pressure to the density at any point is equal to

$$ag / \log \frac{p'}{p}.$$ [M. T., 1893.]

12. A barometer tube consists of three parts whose sections starting from the lowest are A, B, and C. The column consists partly of mercury and partly of glycerine, so that for a certain atmospheric pressure the glycerine just fills that part of the tube whose section is B. Shew that if

$$A : B :: B : C :: 1 : \lambda,$$

and if μ is the ratio of the density of glycerine to that of mercury, the sensitiveness of this barometer is greater than that of a mercury barometer in the ratio $1 : \lambda + \mu - \lambda\mu$, the alteration of level in the cistern being neglected. [M. T., 1881.]

13. If the law connecting the pressure and density of the air were $p = k\rho^2$, prove that, neglecting changes of temperature and gravity, the height of the atmosphere would be twice the height of the homogeneous atmosphere. [Peterhouse, 1886.]

14. If P be the weight of a diving-bell, P' of a mass of water the bulk of which is equal to that of the material of the bell, and W of a mass of water the bulk of which is equal to that of the interior of the bell, prove that, supposing the bell to be too light to sink without force, it will be in a position of unstable equilibrium, if pushed down until the pressure of the enclosed air is to that of the atmosphere as W to $P - P'$. [M. T., 1857.]

15. Two equal vertical cylinders of length l stand side by side, and there is a free communication between their bases. They are partly filled with mercury and a heavy piston is placed at the top of each and allowed to descend slowly: shew that if the difference of the weights of the pistons be greater than the weight of the mercury, air will pass from one cylinder to the other and find the position of equilibrium when the difference of the weights is less than the weight of the mercury. [M. T., 1882.]

16. If a barometer tube dips into a cylindrical cistern of mercury and is suspended by a string which passes over a pulley and supports a counterpoise, prove that the ratio of the changes of height of the counterpoise to the corresponding changes of height of the barometric column is equal to the ratio of the sectional area of the tube to the annular sectional area of the tube.

[M. T., 1875.]

17. A piston of weight W serves as a stopper to confine some gas within a vertical cylinder open at the top. If h is the equilibrium height of the base of the piston above that of the cylinder and if it is slowly displaced through a small distance x, without any change in the temperature of the gas taking place, the work done on the system is $Wx^2/2h$ nearly.

[M. T., 1885.]

18. Two equal vertical tubes are connected at their lower ends by a straight horizontal tube of length a. The upper ends of the tubes are open, and mercury is poured in. The end of one of the tubes is then closed at a height b above the mercury surface, and the system is rotated round the closed tube with angular velocity ω. Prove that, if h be the height of the mercury barometer, the mercury surface in the closed tube will sink through a depth z which is the positive root of the equation

$$4gz^2 + z\{2g(h+2b) - \omega^2 a^2\} - \omega^2 a^2 b = 0.$$

[Trin. Coll., 1889.]

19. If P, P' be the weights which balance a body when weighed in air and in water respectively when the absolute temperature is t, and the corrected height of the barometer is h, prove that the density of the body at the temperature T is

$$\{1 + k(t - T)\}\left(\frac{Pa_t}{P - P'} - \frac{P'b}{P - P'} \cdot \frac{h}{H} \cdot \frac{T}{t}\right),$$

where a_t is the density of water at the temperature t, k is the coefficient of expansion of the body, and b is the density of air when the absolute temperature is T and the corrected height of the barometer is H.

[M. T., 1875.]

ON GASES. 151

20. A barometer tube is filled with n superincumbent liquids of densities $\rho_1, \rho_2 \ldots \rho_n$ in descending order and $A_1, A_2 \ldots A_n$ are the sectional areas of the barometer tube at the upper surfaces of the liquids, A of the open cistern. Prove that the fluctuations of the height of this barometric column are to those of a column of uniform density σ in the ratio

$\sigma (1/A_1 + 1/A) : \rho_1(1/A_1 - 1/A_2) + \rho_2(1/A_2 - 1/A_3) + \ldots \rho_n(1/A_n + 1/A)$.

[M. T., 1883.]

21. A number of cylindrical diving-bells, each formed of uniform thin material whose specific gravity is large compared with that of water, float in water with their mouths downwards and their tops above the surface: prove that if a communication is opened between the air in all the bells they will all sink to the bottom except that bell whose weight is least in comparison with the pressure of the atmosphere on its base, it being assumed that no bell turns over and no air escapes. [M. T., 1888.]

22. In ascending a mountain the temperature of the air is found to decrease by a quantity proportional to the height ascended, and h, k are the observed heights of the barometer at two stations whose difference of altitude is z: shew that z varies as $h^m - k^m$, where m is a certain constant, and where changes of density in the mercury in the barometer are neglected. [M. T., 1882.]

23. A fixed vertical circular tube full of air has within it two diaphragms of weight w_1, w_2 which fit the tube closely, and are originally in contact with one another. They are separated by water being forced into the tube through a small hole which is closed when the weight of water forced in is w_3. Shew that in the position of stable equilibrium the line joining the weight w_1 to the centre of the tube is inclined to the horizon at an angle

$$\tan^{-1} \frac{w_1\gamma + w_2\gamma \cos\gamma + w_3 \sin\gamma}{w_3(1 - \cos\gamma) + w_2\gamma \sin\gamma},$$

where γ is the angle subtended at the centre of the tube by the water. [M. T., 1878.]

24. A long fine tube, open at the lower end and expanding into a large bulb at the upper, is immersed all but the bulb in water, with the bulb just full of air at atmospheric pressure. Shew that if the tube be raised the surface of the water inside will sink below the surface outside and that provided no air escapes, the pressure in the bulb is very nearly a mean proportional between the pressure inside the tube at the level of the water surface outside and the pressure outside at the level of the bulb. (It is assumed that the pressure of the air at all points of the bulb is the same, and that the volume of the tube is small compared with that of the bulb.)

[Jesus Coll., 1891.]

CHAPTER VIII.

HYDROSTATIC MACHINES.

The Siphon.

100. The siphon is an instrument by means of which we can empty a vessel filled with liquid, without moving the vessel.

It consists of a bent tube ABC, open at both ends: it is filled with water or whatever liquid the vessel contains, the ends being temporarily closed: one end A is placed below the surface of the liquid in the vessel to be emptied, and the other C outside the vessel, and below the level of the liquid surface.

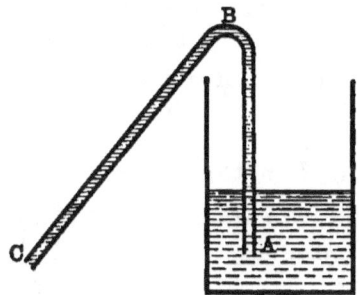

Open the end A and suppose that C is closed by a plug. We shall consider the forces acting on either side of the plug.

Let σ be the density of the liquid in the vessel, h the height of the corresponding barometer, x the depth of C below the surface.

The pressure at the upper side of the plug is $g\sigma h + g\sigma x$ (Art. 23).

The pressure at the lower side is that of the atmosphere, $g\sigma h$.

Hence the force on the upper side exceeds that on the lower, and the plug when free to move is driven out and the liquid after it.

The water flowing out at C tends to cause a partial vacuum at B, in consequence of this the atmospheric pressure on the surface of liquid in the vessel forces the liquid up AB, and there will be a steady flow through the siphon.

We have assumed in the above that the height of B above the surface of the water does not exceed the height of the water barometer, i.e. that B is below the effective surface; otherwise, the water between B and C will flow out at C, and the rest back into the vessel.

It is obvious that the siphon will not work if C be above the water surface, as in that case the force on the lower side of the plug exceeds that on the upper, and the plug is driven *in*.

When once started the siphon is self-acting, the work being done by gravity as the liquid is transferred to a lower level.

101. The velocity with which a liquid flows through a siphon is given by *Torricelli's theorem*.

HYDROSTATIC MACHINES. 155

This theorem gives the velocity of the liquid issuing from a vessel through a small orifice in the side or base, or through a siphon-pipe of narrow bore. It asserts that

If a jet of liquid issue from a vessel through a small orifice, the velocity of the issuing liquid is $\sqrt{2gh}$, where h is the depth of the orifice below the surface of the liquid in the vessel.

Let us first suppose that the liquid issues from a short pipe inserted in the orifice, and fitting it closely. It is obvious in this case and in that of the siphon, that the issuing particles of liquid are moving at right angles to the cross section of the jet.

We shall also assume that the motion is not just beginning and that additional liquid is continually supplied to the surface so that its level remains constant, and the motion is in consequence *steady*.

Let AB be the surface of the liquid in the vessel, and let ab be the cross-section of the jet at the orifice, K being the area of the one and k that of the other.

Let ρ be the density of the liquid.

Let Π be the atmospheric pressure. This is the pressure throughout AB and also throughout ab, since the liquid there has no acceleration in the plane of this section. Let V, v be the velocities at AB, and ab respectively.

Let us consider the motion of the liquid which at a particular instant lies between AB and ab. In an indefinitely short space of time t, the particles originally in AB will sink through a small distance Vt, to the plane $A'B'$, while those in ab have moved along the jet to the section $a'b'$, through the distance vt. As the volume of this liquid remains unchanged, $VK = vk$.

The work done on this liquid is

(1) $\Pi K Vt$, by the thrust of the atmosphere on AB,

(2) $-\Pi kvt$, by the thrust across the section ab,

(3) $gkvt\rho h$ by gravity as the mass $kvt\rho$ is transferred from AB to ab, i.e. through a vertical distance h.

∴ the total work done

$$= \Pi K Vt - \Pi kvt + kvt\rho gh = kvtg\rho h.$$

As the motion is steady, the velocity of the liquid at any point between $A'B'$ and ab remains the same, so that there is no change in the kinetic energy of the liquid between $A'B'$ and ab. The only change in the kinetic energy of the mass under consideration is that instead of the mass $kvt\rho$, with the velocity V at AB, we have an equal mass with the velocity v at ab.

Hence the increase in kinetic energy is

$$\tfrac{1}{2} kvt\rho (v^2 - V^2).$$

But the increase in $K.E.$ = the work done,

$$\therefore \tfrac{1}{2} kvt\rho (v^2 - V^2) = kvt\rho gh,$$

$$\therefore v^2 - V^2 = 2gh,$$

$$\therefore v^2 - v^2 k^2/K^2 = 2gh,$$

$$\therefore v^2 = \frac{2gh}{1 - k^2/K^2}.$$

HYDROSTATIC MACHINES. 157

Since the orifice is small compared with AB, we may neglect k/K, and obtain

$$v^2 = 2gh.$$

If the pipe be turned upwards so that the liquid issues with velocity $\sqrt{2gh}$ vertically upwards, the jet will, but for the resistance of the air, rise to a height h above the orifice, i.e. to the level of the surface in the vessel.

When the liquid issues merely through a hole in the side or base of the vessel, the above reasoning requires modification as we cannot in that case assume that the issuing particles are moving perpendicularly to the cross section of the jet: it is, indeed, an observed fact that the jet contracts for a short distance after leaving the orifice and then expands again. The place where the jet is most contracted is termed the *vena contracta*, and at this place the liquid is moving at right angles to the cross section of the jet. By proceeding as before we find that the velocity at the *vena contracta* is $\sqrt{2gh}$, where h is its depth below the surface in the vessel.

Ex. 1. If m lbs. of fluid issue through a siphon per second, and the amount of fluid in the vessel be kept constant by allowing the supply of fresh fluid to flow gently into it, shew that the apparent weight of the vessel and its contents are less than the actual weight by m lbs. weight provided the orifice of the siphon be 16 feet below the surface level in the vessel, and the water issue vertically downwards.

Ex. 2. A cylindrical vessel with its axis vertical is filled with water and is closed by a heavy piston of mass M. There is a small hole in the piston through which the water escapes. Shew that the uniform velocity of efflux is $\sqrt{2gM/\rho A}$; where ρ is the density of the fluid and A is the area of the cross section of the cylinder. [Trin. Coll., 1889.]

102. The Pumps.

A *common syringe* affords an example of the pump in

its simplest form. It consists of a hollow cylinder AB, ending in a nozzle C. A solid piston works inside the cylinder.

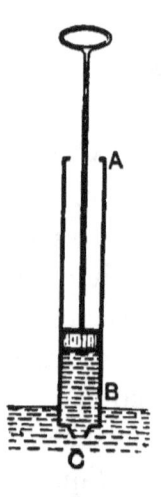

Let the nozzle be held under water, when the piston is at B; now draw the piston back, so that a partial vacuum is formed: in consequence of this the atmospheric pressure forces the water into the cylinder. When the piston is pushed back again, the water is ejected through the nozzle in a jet.

The principle of the above and that of the pumps may be described as that of *suction*. It consists in enlarging a space to which a liquid has access, and consequently creating a partial vacuum. The pressure on the surface of the liquid inside the space is diminished and the atmospheric pressure on the surface outside forces the liquid higher up into the enlarged space. A familiar instance of this is when liquid is sucked into the mouth through a straw. The same principle applies to air as well as to liquids; for instance, to draw air into the lungs, we enlarge them by raising the walls of the chest.

103. *Valves* are used in the construction of many of the hydrostatic machines. A valve is a contrivance which allows water, air, or steam to pass through it in one direction, but not in the other. A very simple form of valve is the leather disc which closes the opening on the lower side of an ordinary pair of bellows. When the bellows are expanded the pressure below the flap opens it; when they are compressed the pressure inside shuts the flap down against the edge of the opening and prevents the air from escaping that way. It is a form of the *hanging flap* valve, which is

HYDROSTATIC MACHINES.

generally a metal disc turning on a hinge. Another valve is the *ball clack* valve, which consists of a metal ball, fitting into and closing a circular orifice, but lifting when the excess of pressure is below. The ideal valve opens with the smallest excess of pressure on one side and permits no leakage the other way, but practically in all valves a definite excess of pressure is required, and a certain amount of leakage occurs.

104. The Common Pump consists of a cylinder AB, with a spout E opening out of the upper part; at the bottom is a valve C opening out of the pipe CD which communicates with the water to be raised. A piston P with a valve F opening upwards works between B and E.

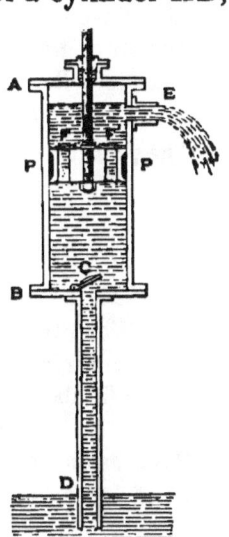

(In the figure there are two piston valves F and F'.)

The action is as follows:—

Suppose the piston to be at B initially: when it is raised, a vacuum is created in PB, the atmospheric pressure consequently closes the piston valve, and the air in CD opens C and fills PB; this reduces the pressure in the pipe and the water rises in it. When the piston is pushed down, the air in PB is compressed and in consequence the increased pressure closes the valve C and opens that in the piston, so that the air escapes through the piston valve. After this is repeated several times, the water rises into the cylinder and is forced through the piston valve when the piston is lowered, and lifted out through E when it is raised again.

105. *To find the height the water rises in one stroke of the piston,* (*the* n+1*th*).

Let P, Q be the water-surfaces at the beginning and end respectively of the upward stroke, y, z their heights above the surface outside. Let B be the valve between the pipe and the cylinder, c its height above the surface of the water outside. Let h be the height of the water barometer, d the length BD of the stroke of the piston, a, β the areas of the sections of the cylinder and pipe, respectively.

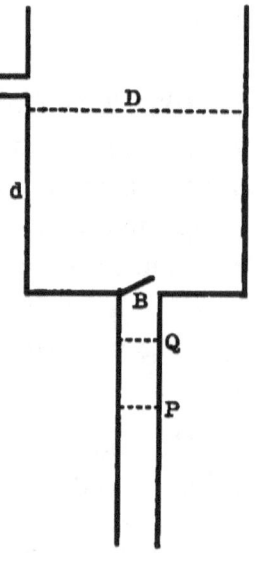

(i) When Q is *below* B.

The air originally of volume PB or $(c-y)\beta$ expands into the volume $QB + BD$,
$$(c-z)\beta + ad.$$
The pressure alters in the ratio of
$$h-y : h-z,$$
∴ by Boyle's law,
$$(h-y)(c-y)\beta = (h-z)\{(c-z)\beta + ad\}.$$
This equation determines z, when y is known.

(ii) When Q is *above* B.

The air originally of volume PB or $(c-y)\beta$, expands into the volume $BD - QB$, i.e.
$$a\{d-(z-c)\}.$$
Hence, as before, we have
$$(h-y)(c-y)\beta = (h-z)(d-z+c)a.$$

106. The Lift Pump consists of a cylinder AB, with a valve C at the bottom opening from the pipe CD, which communicates with the water to be raised: there is also a valve E at the top opening into a pipe EF, up which the water is to be lifted. In the cylinder, a piston P works through an air-tight collar between A and B. There is a valve G in the piston opening upwards.

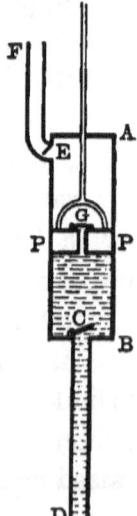

The mode of working is as follows:—

Suppose the piston to be initially at B.

When the piston is raised a partial vacuum is created in PB, and in consequence the air in CD opens the valve C and enters PB, the water rising in the pipe: also the compressed air in AP closes the piston valve and opening the valve E escapes up EF. When the piston is pushed down again, a vacuum is created in PA and the air in PB is compressed. This closes the valves C and E, but opens the piston valve so that the air in PB enters PA. After several strokes of the piston the water passes through C, on the next downward stroke it passes through the piston-valve, and in the next upward stroke is lifted up EF through E.

107. The Force Pump consists of a cylinder AB, with a valve F at the bottom opening from a pipe BD, which communicates with the water. Very near the bottom of AB there is a valve C opening into a pipe CE, up which the water is to be forced. A solid piston works between A and B.

(Instead of a solid piston, a pole H working in an air-tight collar is sometimes used, as in the figure, and the pump is termed a *plunger pole pump*.)

Suppose the piston to be at B initially: when it is raised, the pressure in AB is diminished, and in consequence the air in BD opens F and enters AB, while the water rises in BD; at the same time the atmospheric pressure closes the valve C. When the piston is pushed down again, the air in AB is compressed, and consequently closes the valve F and ultimately opens the valve C and escapes up

CE. When several strokes of the piston have been made, the water rises up through *F* and, on the next downward stroke, is forced up *CE*.

Ex. If A be the area of the section of the piston of a force pump, l the length of the stroke, n the number of strokes per minute, B the area of the pipe from the pump, find the mean velocity with which the water rushes out. [M. T., 1860.]

108. In all the three pumps above described the water rises in the pipe below the cylinder through atmospheric pressure and consequently none of them will work when the valve at the bottom of the cylinder is more than 32 feet, the height of the water barometer, above the water level outside: as a matter of fact, owing to the imperfection of the valves, it is found that the maximum height is several feet less than this. In the lift and force pumps the limit to the height the water can be forced up the pipe leading out of the cylinder depends only on the strength of the apparatus and the force that can be applied to the piston.

109. In order that the force or lift pump may throw up a continuous stream of water, an air-chamber is introduced. The valve *C* from the cylinder opens into a strong chamber *A*, out of the lower part of which a vertical pipe *BD* leads. The water is first forced into the chamber, and then up the pipe; the air above the level of the lower end *B* of the pipe cannot escape. When the piston is descending quickly,

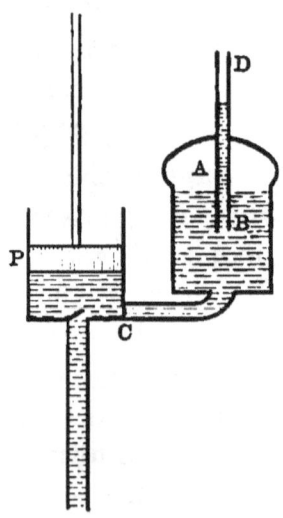

which will be about the middle of the stroke, the water will be forced into the air-chamber A with considerable velocity, and the air will be compressed; during the upward stroke, the air will expand again and continue to force the water up the pipe. The energy which is stored up in the compressed air during the downward stroke is consumed during the upward stroke. The same principle applies to organ or forge bellows: a weight is raised during one period and descends during another, and so the air is forced out continuously.

110. The Fire Engine consists of a double force pump, provided with an air-chamber as above. The two pistons are attached to the opposite ends of a lever, so that one descends while the other ascends.

111. Bramah's or the Hydraulic Press consists of

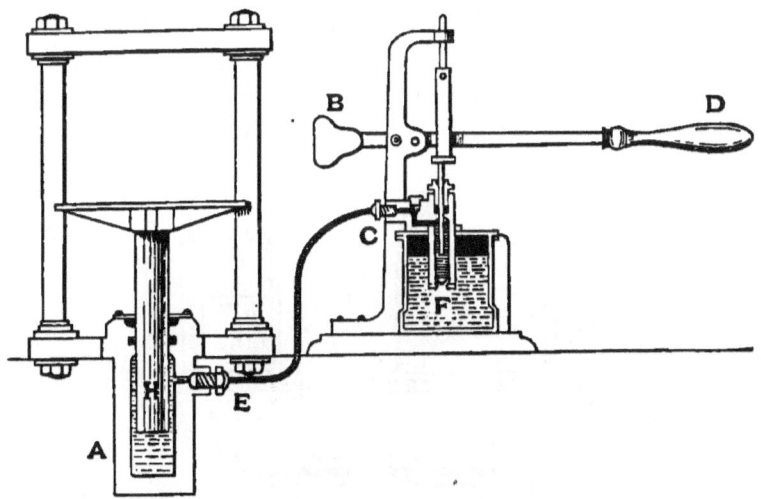

a force pump the pipe from which CE opens into a strong cylinder A, in which H, a large solid cylinder, works through a water-tight collar. The piston of the pump

164 HYDROSTATIC MACHINES.

is a plunger pole much smaller in section than H and working through a water-tight collar; it is moved by a lever BD. When the piston is forced down, the valve opening into CE is opened, and by the principle of the transmissibility of fluid pressure the pressure everywhere is the same, so that if the section of the solid cylinder be 100 times that of the plunger pole, the upward force on the former is 100 times the downward force on the latter.

Hawksbee's Air-Pump.

112. This consists of a cylinder AB called the *barrel*, with a valve B at one end opening outwards from a pipe CD leading to the vessel E to be exhausted called the *receiver*. A piston P, with a valve F in it opening outwards, works in the barrel.

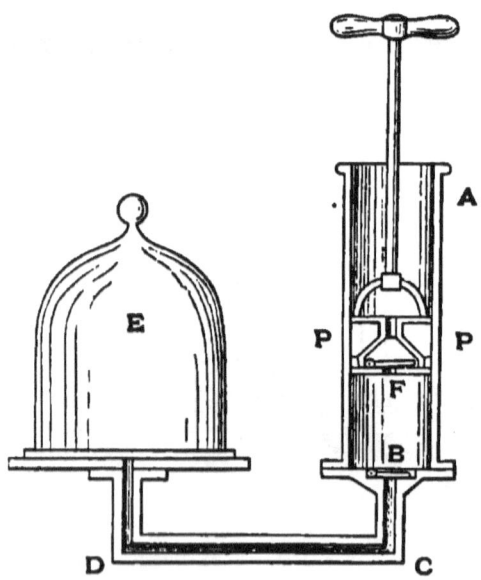

In the original pump made by Hawksbee there were two barrels, and the pistons were so connected that when one was going up the other was going down.

HYDROSTATIC MACHINES. 165

The mode of working is as follows :—

Suppose the piston to be initially at B. When the piston is withdrawn, a partial vacuum is created in PB, so that the atmospheric pressure closes the piston valve and the air in the receiver opens the valve at B and expands into the barrel. When the piston is pushed back again, the air in PB is compressed, the valve B is closed, and ultimately the piston valve is opened, and the air in PB escapes into the atmosphere. In this way at every complete stroke a barrelful of air is removed from the receiver. The exhaustion is however never complete, as the valve B does not open when the pressure in the receiver is less than a certain amount: also as there is always some 'clearance' space between the piston and the end of the barrel, the pressure in PB will not always be able to open the piston valve.

Smeaton's Air-Pump.

113. This is the same as Hawksbee's, except that the end A of the barrel is closed, the piston works through an air-tight collar, and there is a valve at A opening outwards.

The method of working is similar to that in Hawksbee's but it has the advantages of carrying the exhaustion further, and of being easier to work. The first advantage is due to the pressure in PA, as the piston descends, being less than the atmospheric, so that the piston valve has a better chance of opening than in Hawksbee's. The second advantage is due to the pressure in PA being less than the atmospheric during the greater part of the upward stroke, so that less effort is required to raise the piston.

HYDROSTATIC MACHINES.

PROP. *To determine the density of the air in the receiver after* n *strokes of the piston.*

Let V be the volume of the receiver, and v that of the barrel, let ρ be the density of the atmospheric air, and $\rho_1, \rho_2, \rho_3 \ldots \rho_n$ the density of the air in the receiver, after $1, 2, \ldots n$ complete strokes.

After one complete stroke the air which occupied a volume V occupies a volume $V + v$,

$$\therefore \rho_1(V+v) = \rho V.$$

Similarly $\quad \rho_2(V+v) = \rho_1 V,$

and $\quad \ldots\ldots\ldots\ldots = \ldots\ldots$

$$\rho_n(V+v) = \rho_{n-1} V,$$

$$\therefore \rho_n(V+v)^n = \rho V^n.$$

Ex. 1. Supposing the upper valve of Smeaton's air-pump to open when the piston is three-quarters of the way up, what was the density of the air in the receiver at the beginning of the ascent? [Jesus Coll., 1886.]

Ex. 2. If the piston works only between two points distant a and b respectively from the top and bottom of the barrel, and at a distance c apart, while the receiver is equal to a length l of the barrel; find the density of the air in the receiver after two strokes of the piston.

114. Tate's Air-Pump consists of a barrel AB, with an opening C half-way down leading to the receiver. Two solid pistons P and Q rigidly connected together are moved up and down by a rod working through an air-tight collar at D. At the top and bottom of the barrel are the valves F and F', which open outwards. As the piston is pushed down the air in CB is driven out through F', and as it is raised the air in CA is driven out through

F, the air always flowing in from the receiver through C to refill the barrel.

This pump combines the advantages of Smeaton's with the additional one of producing a greater degree of exhaustion, due to the piston-valve and that leading to the receiver being dispensed with.

115. Sprengel's Air-Pump consists of a long straight vertical glass tube BC. This communicates by means of a short flexible tube A with a vessel E above it. A can be closed by a clamp F. At B a tube BD leads out of ABC and communicates with the receiver. BC dips into a small cup G provided with a spout. The vessel E is filled with mercury, which flows down AB in a continuous stream, but as it passes B the mercury breaks off into drops separated from one another by bubbles of air from the receiver. The vessel E is not allowed to become empty, the mercury being poured back into it. As this process goes on, the mercury rises in CB owing to the diminished pressure in the receiver, and after a time the completion of the exhaustion is announced by the metallic sound made by the falling mercury. The degree of exhaustion is that of the Torricellian vacuum—this is shewn by the column of mercury in BC being ultimately equal to the barometric height.

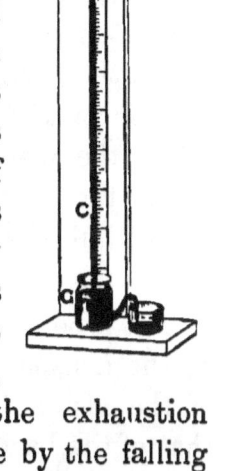

It is of course necessary that BC should be greater than the barometric height, otherwise the mercury will flow into the receiver.

The Condenser.

116. This is an instrument for compressing air and is the reverse of the air-pump. It consists of a hollow cylinder, AB, the barrel, from which a valve F opens into a strong receiver C. In the barrel a piston P with a valve E opening inwards works. There is generally a stop-cock Q between B and the receiver.

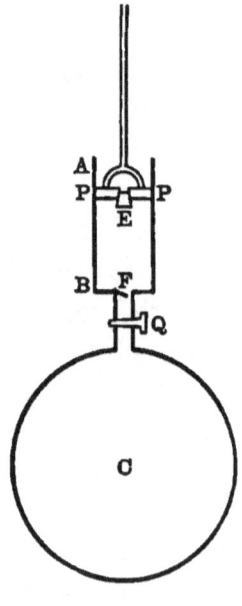

Suppose the receiver to be full of air at atmospheric pressure and the piston to be at B. When the piston is withdrawn a vacuum is created in PB, in consequence of which the valve F closes, and the atmospheric air opens the piston valve E and fills the barrel. When the piston is pushed down again the air in PB is compressed, and so closes the piston valve, opens F and enters the receiver. At every stroke a barrelful of air at atmospheric pressure is forced into the receiver.

Let ρ be the density of the air at atmospheric pressure, ρ_n the density of the air in the receiver after n strokes of the piston. Let V be the volume of the receiver, v that of the barrel.

After n strokes the total volume of air at atmospheric pressure in the receiver is $V + nv$,

$$\therefore \rho(V + nv) = \rho_n V,$$
$$\therefore \rho_n = \rho(1 + nv/V).$$

In some condensers the piston has no valve, the valve opening into the receiver is at the side of the barrel near

HYDROSTATIC MACHINES. 169

the bottom, and at the bottom of the barrel is the valve opening from the atmosphere.

Ex. 1. If of the volume B of the cylinder only C is traversed by the piston, prove that the pressure in the receiver of a condenser cannot be made to exceed $B/(B-C)$ atmospheres. [Jesus Coll., 1890.]

Ex. 2. In a certain condenser, the area of the piston-section is 5 square inches, and the volume of the receiver is ten times the volume of the range of the piston. If the greatest force which can be used to make the piston move is 165 lbs., find the greatest number of complete strokes which can be made. [St John's Coll., 1881.]

Gauges.

117. Gauges for measuring pressures less than the atmospheric pressure, as for instance the pressure inside the receiver of an air-pump, are termed *vacuum-gauges*. The *barometer* and *siphon gauges* are gauges of this kind.

The *barometer gauge* consists of a vertical glass tube, open at both ends, the upper end being in communication with the chamber the pressure in which is required, and the lower end dipping into some mercury in a vessel. The height of the mercury surface in the tube above that in the trough measures the excess of the atmospheric pressure above the pressure in the chamber.

A modification of this gauge is the *open tube vacuum gauge*. This is a U-tube open at both ends, and containing mercury. One branch communicates with the receiver of the air-pump, and the other with the atmosphere.

118. The *siphon* or *closed tube vacuum gauge* consists of a U-tube, containing mercury, closed at one end and open at the other. At the closed end is a Torricellian vacuum and the open end communicates with the receiver of an air-pump. The height of the mercury surface in the closed branch above that in the open branch measures the pressure in the receiver. In using this gauge it is not necessary to know the atmospheric pressure.

119. Gauges used to determine pressures greater than that of the atmosphere, as for instance the pressure inside the receiver of a condenser or in the boiler of a steam-engine, are often termed *manometers*.

The open tube method can obviously be applied to the case of pressures not very much greater than the atmospheric pressure. It is however very inconvenient for the purpose of determining large pressures.

The *condenser gauge* is a uniform glass tube AB, closed at the end B, but open at A and communicating with the receiver of the condenser. At the end B is

some air separated from the air in the condenser by a drop of mercury. When the mercury is in equilibrium, the pressure is the same on either side of it. Let C be the position of the drop when the pressure inside the receiver is that of the atmosphere. Then when the drop is at D the pressure in the receiver is $\Pi \cdot BC/BD$ (Boyle's Law), Π being the atmospheric pressure.

120. The *compressed-air manometer*. A particular form of this consists of a U-tube ABC, containing mercury. The end C is open and communicates with the chamber the pressure in which is required: the other end A is closed and contains some dry air. When the pressure in the chamber is that

HYDROSTATIC MACHINES. 171

of the atmosphere the mercury surfaces P, Q in the two branches are on a level. As the pressure in the chamber increases, the mercury surface in the left-hand branch is pushed up to some point P' and the air there compressed, while that on the other side is pushed down to Q'.

The pressure in AP' is deduced from the change in volume, and that in CQ' exceeds that in AP' by an amount proportional to the difference of level between P' and Q'.

In some forms of this manometer the section of the tube varies considerably.

ILLUSTRATIVE EXAMPLE.

If h be the range of the piston in Smeaton's air-pump, a its distance from the top of the barrel in its highest position, b its distance from the bottom in its lowest position, and ρ the density of the atmosphere: prove that the limiting density of the air in the receiver will be

$$\frac{ab}{(h+a)(h+b)} \rho.$$

[M. T., 1861.]

Let A be the upper valve, B the lower, and P the piston-valve. It is quite clear that the limiting density of the air in the receiver is not reached when either A or B opens as the piston rises. When the piston is in its lowest position, A and B are shut and P is open, so that the density throughout the barrel is constant; let this density be ρ_1. Let ρ' be the density in the receiver. As the piston rises, B will open if ρ' is greater than the density in the barrel below the piston, when the piston is in its highest position, i.e. if ρ' is greater than $\rho_1 \cdot \dfrac{b}{h+b}$.

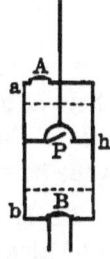

Also, A will open if the density above the piston when in its highest position is greater than ρ, i.e. if $\rho_1 \cdot \dfrac{h+a}{a}$ is $> \rho$.

Hence either A or B will open if either

$$\rho' \text{ is} > \rho_1 \cdot \frac{b}{h+b},$$

or
$$\rho_1 \text{ is} > \rho \cdot \frac{a}{h+a},$$

i.e. if
$$\rho' \text{ is} > \rho \cdot \frac{ab}{(h+a)(h+b)}.$$

Hence so long as ρ' exceeds this value the exhaustion proceeds, but no longer.

EXAMPLES. CHAPTER VIII.

1. Prove that if the area of the section of the *vena contracta* of a jet of liquid of density ρ be A, the amounts of liquid and momentum which issue in a time t are $At\sqrt{2p\rho}$, and $2Apt$ respectively, where p is the amount by which the pressure would be increased at the orifice, if the liquid were at rest.

2. A siphon with vertical arms filled with mercury (ρ) and closed at both ends is inserted in a basin of water (σ). Prove that when the stoppers are removed, provided the barometer is high enough (1) if k, the whole length of the outside arm, be $> h$, the whole length of the inside arm, the mercury will flow out followed by water: (2) if h be $> k$, the end of the immersed tube must be at a depth below the free surface $> (h-k)\rho/\sigma$, in order that the mercury may not flow back into the basin.

3. Prove that in the common pump the water will just rise into the upper cylinder at the end of the second stroke if

$$H^2\left(1 - \frac{a}{nb}\right)\left(2 - \frac{a}{nb}\right) - H\left(4a + nb - \frac{3a^2}{nb}\right) + a(2a + nb) = 0,$$

where a, b are the lengths of the lower and upper cylinders, n is the ratio of the sectional area of the latter to that of the former, and H is the height of the water barometer. [M. T., 1889.]

4. In Smeaton's air-pump, find the position of the piston in its $(n+1)$th ascent when the highest valve begins to open: and shew that in that position the tension of the piston rod : thrust of the atmosphere on the piston :: $1 - \{A/(A+B)\}^n : 1 - \{A/(A+B)\}^n \cdot B/(A+B)$, A and B being the volumes of receiver and cylinder.

[M. T., 1856.]

HYDROSTATIC MACHINES.

5. Prove that if the piston of Hawksbee's air-pump cannot traverse the whole length of the cylinder, the density in the receiver after n strokes will be

$$1-\left\{1-\left(\frac{A}{A+B}\right)^n\right\}\frac{C}{B}$$

of the density of the atmosphere, supposing A to denote the volume of the receiver, B that of the cylinder, and C that of the part traversed by the piston. [M. T., 1883.]

6. In a condenser the motion of the piston is checked by permanent stops at heights h, k from the receiver-valve A. Find the relation between the heights of the piston above A at which the piston-valve and the valve A open in any upward and in the next downward stroke respectively. [M. T., 1885.]

7. If the greatest and least volumes between the valves of a condenser be V and v and if the valves be open or closed according as the difference of pressure on the two sides be greater or less than p, shew that the limiting pressure in the receiver will be

$$(\Pi-p)V/v-p,$$

where Π is the pressure of the atmosphere. [M. T., 1872.]

8. If there is a leak in the receiver of a condenser, and the volume of air measured at atmospheric pressure expelled through it in a minute be V multiplied by the difference of the logarithms of the pressures at each side of the leak: prove that the maximum density attainable is n times that of the atmosphere where

$$\log n = sA/V,$$

s being the number of strokes per minute, and A the capacity of the barrel of the pump. [M. T., 1870.]

9. Prove that if the density of the air in a receiver of volume V be increased tenfold, energy amounting to at least $\Pi V(10\log_e 10 - 9)$ must be expended, where Π is the atmospheric pressure and the temperature is supposed uniform throughout.

[Jesus Coll., 1892.]

174   HYDROSTATIC MACHINES.

10. A small hole is made in a partition separating two reservoirs of unlimited extent, at the depth h below the surface of the water which stands highest, and the depth k below the other surface: shew that the water will flow through the hole with velocity

$$\sqrt{2g}\,(\sqrt{h}-\sqrt{k}),\ \text{nearly.}$$

[Smith's Prizes, 1861.]

11. Inside a condenser is placed a mercury barometer; shew that after n strokes of the piston the column will have risen a distance approximately equal to

$$n\frac{Bh}{R}\frac{K}{K+h}\left\{1-\frac{Kkh}{K+k}\frac{R+nB}{R^2}\right\},$$

where h is the initial height of the mercury column, k the area of its cross section, K that of the mercury in the basin, B is the volume of the barrel, and R that of the receiver initially, which is large compared with the volume occupied by the mercury.

[Trin. Coll., 1891.]

CHAPTER IX.

CAPILLARITY.

121. It has been already mentioned (Art. 7) that though the surface of a considerable mass of liquid in equilibrium is horizontal, this is not necessarily the case if the amount of liquid is small. A small drop of mercury for instance, when placed on a horizontal glass plate, does not flatten itself out, and if it is forcibly flattened out, it resumes its old shape when released, in opposition to the action of gravity. Also, if a glass tube of fine bore open at both ends be partially immersed in water, the water rises in it to a certain extent, so that the surface inside the tube is above that outside. These facts seem at variance with the result obtained in Art. 22.

To reconcile theory with observation we are driven to the conclusion that there must be in action forces which we have not hitherto considered, or in other words, that the potential energy of a liquid does not depend entirely on its position relative to the earth, but also partly on its own configuration.

122. Very important conclusions may be drawn from the following experiment devised by Plateau. He placed a quantity of oil in a mixture of alcohol and water of the same density as the oil, and found that the oil when it was free to do so took a spherical shape: also that it resumed this shape if it was distorted and then released. The potential energy of the mass must therefore be least when the oil has its spherical shape, i.e. when the area of its surface is a minimum. Since the density of the whole mass is uniform the shape the oil assumes cannot affect the part of the energy depending on gravity. We conclude therefore that part of the energy of the system depends on the position of the particles relative to one another, independently of their position relative to the earth. Also as the energy is least when the surface is least, a particle very near the surface must have more energy than one at some distance from it. If we assume that it is only very near the surface where this increase in energy occurs, it follows that the energy due to some of the particles being near the surface is proportional to the area of the surface.

Under this assumption the energy of the mass consists of three parts. One part, due to gravity, is equal to the product of the weight and the height of the centre of gravity above the earth's surface; another, due to molecular forces, is proportional to the volume and independent of the shape; the third part, the surface-energy, is due to the excess of the energy of a particle near the surface above that of one distant from it, and is proportional to the area of the surface.

123. By coating the sides of a vessel containing mercury with very thin films of different solid substances, Quincke was able to obtain limits within which the molecular forces, to which the surface-energy is due, are sensible. He found that for sulphide of silver the capillary phenomena were independent of the thickness of the film so long as it exceeded ·000048 mm. and for iodide of silver, so long as it exceeded ·000059 mm. We infer then that it is only within these limits the molecular forces are sensible.

124. The general principle, from which capillary phenomena may be deduced, may be stated as follows:

When two fluids which do not mix, or a solid and a fluid, are in contact, a portion of their energy is equal to the product of the area of the surface separating them and a quantity which is constant for the same pair of substances and the same temperature, but which varies for different pairs of substances. This quantity for any pair of substances may be called their mutual *surface-energy per unit area.*

If we assume that the energy of a unit area of a mercury-glass surface is greater than that of a unit area of an air-glass surface we can explain why, when a fine glass tube is dipped into mercury, the mercury inside the tube is depressed. Since the energy of a material system always tends to a minimum, the mercury-glass surface tends to decrease and the air-glass surface to increase, and this tendency produces the observed effect in opposition to gravity. Also a drop of mercury does not spread itself out indefinitely on a horizontal plane, because that would involve an indefinite increase of the energy of the surfaces separating the mercury from the air and from the plane.

A drop of oil placed on the surface of some water spreads out indefinitely or until it covers the whole surface, because the energy per unit area of the water-air surface is greater than that of the oil-air surface and that of the oil-water surface together.

125. When three fluids which do not mix are in equilibrium with the surfaces separating each pair intersecting in a common line, it is a well-known experimental fact that the angles between the different surfaces are constant for the same three fluids. This fact can be deduced from the theory of surface-energy.

Let A, B, C be the three fluids; let S_{bc}, S_{ca}, S_{ab} be the surface-energies per unit area of the surfaces separating B and C, C and A, A and B respectively. Let O be the point where the plane of the paper cuts at right angles the common line of intersection of the three surfaces; let OP, OQ, OR be its intersections with the three surfaces, α, β, γ the angles between them respectively.

Since the fluids are in equilibrium, if an indefinitely small virtual displacement be given to the system, the total alteration in energy is zero.

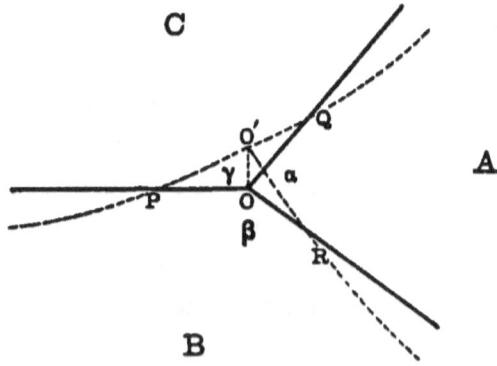

Let the common line of intersection be displaced parallel to itself and in a direction perpendicular to the surface (BC), so that O' is its intersection with the plane of the paper and OO' (x) is indefinitely small.

The increase in energy of the surface (AB) per unit length of the common line of intersection is

$$S_{ab} \cdot x \cos OO'R = S_{ab} \cdot x \sin \beta.$$

Similarly that for the surface (AC)

$$= - S_{ca} \cdot x \sin \gamma.$$

That for the surface $(BC) = 0$.

The alteration in the energy due to gravity (or any external force) will be a small quantity of a higher order than x, as it will depend on the indefinitely small displacements of indefinitely small volumes.

Hence $\quad S_{ab} \sin \beta = S_{ca} \sin \gamma,$

$$\therefore \frac{S_{ab}}{\sin \gamma} = \frac{S_{ca}}{\sin \beta} = \frac{S_{bc}}{\sin \alpha}, \text{ similarly.}$$

As the three angles α, β, γ, together make up four right angles, it follows from these equations that they are the external angles of a triangle whose sides are proportional to S_{bc}, S_{ca}, S_{ab}, respectively.

Hence the angles α, β, γ are constant for the same three fluids.

126. When two fluids which do not mix (water and air for instance) have a surface of separation which meets the surface of a solid (glass for instance), it can be shewn that the angle between the surface of the solid and the separating surface is constant.

180 CAPILLARITY.

If we regard the solid as the substance C, and the two fluids as A and B, by displacing the line of intersection parallel to itself along the surface of the solid through

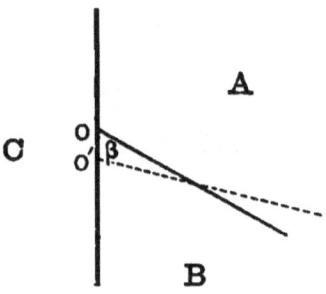

an indefinitely small distance, we obtain from the equation of virtual work,

$$S_{bc} + S_{ab} \cos \beta - S_{ac} = 0,$$

$$\therefore \cos \beta = \frac{S_{ac} - S_{bc}}{S_{ab}},$$

i.e. β is constant, which agrees with the observed results.

Rise of liquid between two glass plates.

127. If two vertical glass plates are placed parallel to one another and at a short distance apart in a vessel containing water, it is found that the surface of the water between them is at a higher level than that outside. We can determine this difference of level in terms of the constant angle β, which the air-water surface makes with the plates, the surface energy of the air-water surface, and the distance between the plates.

(It is assumed that the distance between the plates is so small compared with that between the sides of the vessel, that a small rise or fall of the surface between

CAPILLARITY.

the plates produces no appreciable change in the level outside.)

Let S_{bc}, S_{ca}, S_{ab} be the surface-energies per unit area of water-glass, air-glass, and air-water surfaces respectively. Let w be the weight per unit volume of water, h the difference of level of the air-water surfaces, and d the distance between the plates.

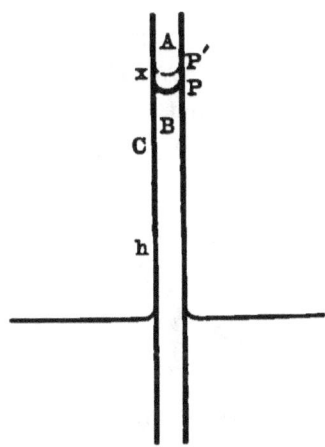

Let the air-water surface P between the plates be raised to P' through a small distance x. In this displacement the work done against gravity per unit length (measured horizontally) of the plates is $whdx$, since a volume dx of water is raised from the level of the surface outside to that inside.

The increase in the water-glass surface per unit length of the plate = the decrease in the air-glass surface = $2x$, ∴ the decrease in surface-energy per unit length of the plate
$$= 2\,(S_{ac} - S_{bc})\,x.$$

Hence the equation of virtual work is

$$2(S_{ac} - S_{bc})x - whdx = 0,$$

$$\therefore whd = 2(S_{ac} - S_{bc}) = 2S_{ab}\cos\beta. \quad \text{(Art. 126.)}$$

Hence the height the water rises is inversely proportional to the distance between the plates.

*128. If one of the plates be free it can be shewn that it will move towards the other.

The energy, E, of the system per unit length of the plates is the sum of a constant quantity together with a quantity a, depending on the rise of the liquid between the plates above the level outside, and a quantity b, depending on the area of the plates touched by air and liquid respectively.

$$a = whd \cdot \frac{h}{2},$$

$$b = 2(S_{bc} - S_{ac})h.$$

Hence E

$$= \text{constant} + \tfrac{1}{2}wh^2d + 2h(S_{bc} - S_{ac})$$

$$= \text{constant} + \frac{2(S_{ab}\cos\beta)^2}{wd} - \frac{4(S_{ab}\cos\beta)^2}{wd} \quad \text{(Art. 127),}$$

$$= \text{constant} - \frac{2(S_{ab}\cos\beta)^2}{wd}, \text{ i.e. it diminishes with } d.$$

Since the system always tends to take up the position in which its energy is least, the plates will move towards one another, if free to move. Let P be the force per unit length which must be applied to each plate to keep it in equilibrium.

If d be reduced to $d-x$, where x is indefinitely small, the decrease in E

$$= \frac{2(S_{ab}\cos\beta)^2}{w} \cdot \frac{x}{d(d-x)} = \frac{2(S_{ab}\cos\beta)^2 x}{wd^2}$$

$$= \tfrac{1}{2}wh^2x.$$

The work done by $P = -Px$.

Therefore since the total work done during this displacement is zero

$$\tfrac{1}{2}wh^2x - Px = 0,$$

$$\therefore P = \tfrac{1}{2}wh^2.$$

129. If instead of two plates, a glass tube of fine bore (internal radius r) is placed vertically in water, we can obtain the height, h, the water will rise in it.

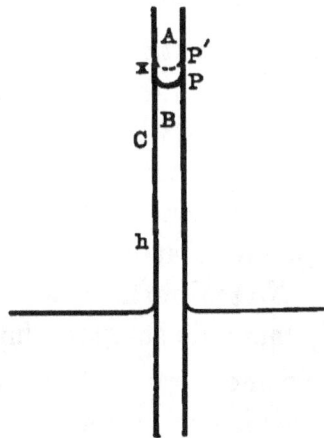

With the notation of Art. 127, if the water surface in the tube be raised an indefinitely small distance x, the decrease in surface-energy

$$= 2\pi rx(S_{ac} - S_{bc}).$$

The work done against gravity in raising the volume $\pi r^2 x$ of water from the level outside to that inside is $\pi r^2 xwh$,

∴ the equation of virtual work is

$$2\pi r x \, (S_{ac} - S_{bc}) - \pi r^2 xwh = 0.$$
$$\therefore whr = 2\,(S_{ac} - S_{bc}),$$
$$= 2 S_{ab} \cos \beta, \; (\text{Art. 126})$$

or h varies inversely as r.

130. If mercury, or any liquid which does not *wet* glass, be employed instead of water, the formulae of Arts. 127—129 apply, but as the angle β in that case is obtuse, h is negative, i.e. the mercury surface inside the plates or the tube is depressed below that outside.

Ex. If the capillary rise of water in a glass tube ·05 in. in diameter be ·96 inches, find that between two vertical parallel plates ·02 in. apart.

Flexible Membranes.

131. Just as a flexible string is one which will yield to any force however small tending to bend it, so a flexible membrane or surface is one which cannot resist the action of any force however small tending to bend it. From this it follows that the direction of the stress across any line drawn in the membrane is at every point in the tangent-plane at the point. No real membrane fulfils this condition perfectly, but very many do so approximately.

In some flexible membranes, the stress across a line drawn in the surface is everywhere perpendicular to that line, i.e. there is no shearing stress; in others, the stress may be partly or entirely along the line. In the former case the tension at any point across a line drawn through the point is measured by the limit of T/l, where T is the pull across an indefinitely small length l of the line.

CAPILLARITY. 185

132. Prop. *In a flexible membrane, incapable of exerting a shearing stress, the tension at any point is the same in all directions.*

Through any point A draw two lines AB, AC in the surface at right angles to one another, and join BC.

Let t, t', and t'' be the average tensions across the lines AB, AC and BC respectively. Let w be the external force per unit area on the triangle ABC of the membrane,

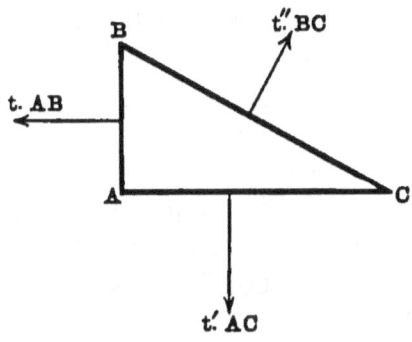

making some angle θ with AC.

Resolving along CA for the equilibrium of the triangle ABC, we have

$$t \cdot AB - t'' \cdot BC \cos B + w \cdot \tfrac{1}{2} AB \cdot AC \cos \theta = 0,$$

$$\therefore t = t'' - \tfrac{1}{2} w AC \cos \theta,$$

∴ when AB, AC are diminished indefinitely,

$$t = t'' \text{ and similarly } t' = t'',$$

i.e. the tensions at A in directions perpendicular to AB, AC and BC are equal.

In the following propositions (Arts. 133—136) it is assumed that the membrane is so light that its weight may be neglected.

*133. Prop. *In a flexible membrane, incapable of exerting a shearing stress, and in the form of a right circular cylinder,* pr = t, *where* r *is the radius,* t *the tension at any point, and* p *the excess of the internal pressure at the point over the external.*

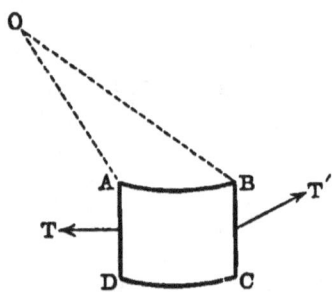

Let $ABCD$ be a small rectangular portion of the membrane, where AD, BC are portions of generating lines, and AB, CD are arcs of circular sections. Let O be the centre of the arc AB. Let T, T' be the pulls across AD, BC, respectively.

The thrust on $ABCD = p \cdot AB \cdot AD$, and makes an indefinitely small angle θ with OA.

Resolving along OA, we have

$$p \cdot AB \cdot AD \cos \theta = T' \sin AOB = T' \cdot \frac{AB}{r},$$

$$\therefore \frac{T'}{AD} = rp \cos \theta,$$

$$\therefore t' = pr \text{ ultimately.}$$

Resolving perpendicular to AO,

$$p \cdot AB \cdot AD \sin \theta + T' \cos AOB = T$$

$$\therefore \frac{T}{AD} - \frac{T' \cos AOB}{AD} = p \cdot AB \sin \theta$$

$$\therefore t - t' = 0 \text{ ultimately,}$$

$$\therefore t = t' = pr.$$

CAPILLARITY. 187

Hence the tension and p are constant throughout the section.

Similarly by resolving along a generating line, it can be shewn that the tension at every point along it is the same, and therefore throughout the membrane.

If the right section of the cylinder be not a circle, the above formulae can be proved in the same way, provided r denotes the radius of curvature at any point of the section. In this case pr is constant throughout the membrane.

*134. If the pressure be the same at all points of a generating line, the reasoning of the last Article will apply to a flexible membrane which can exert a shearing stress, as from symmetrical considerations there will clearly be none across the lines in the figure.

*135. As a particular case of Art. 133, let us consider the shape assumed by a thin rectangular membrane, two of whose sides are attached to the horizontal edges of a box, while the other two sides fit the sides of the box so closely that liquid can be poured above the membrane without escaping.

It is obvious that the membrane will form a cylindrical surface whose generators are horizontal.

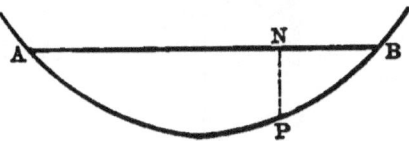

Let APB be a section of this cylindrical surface, AB being in the horizontal surface of the liquid.

Draw PN perpendicular to AB.

The pressure at $P = w \cdot PN$,

t the tension is constant throughout the membrane (Art. 134), and $= pr = wr \cdot PN$, where r is the radius of curvature at P.

∴ the curvature at P is proportional to the depth below the free surface.

The curve APB is termed the *Lintearia*, and is the same as the *Capillary Curve* (Art. 139).

136. Prop. *To prove that in a flexible spherical membrane incapable of exerting a shearing stress, $2t = pr$, where r is the radius, t the tension at any point, and p the excess of internal over external pressure there.*

Let $ABCD$ be a small square-shaped portion of the membrane bounded by great circles.

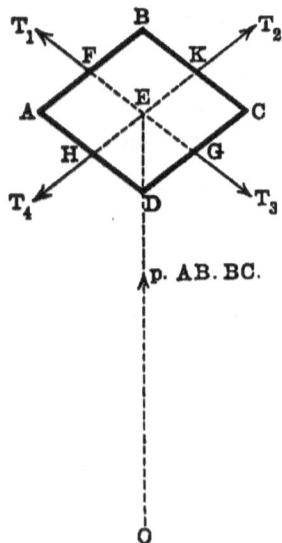

Let FEG, HEK be arcs of great circles bisecting AB, CD, and BC, AD, respectively.

CAPILLARITY. 189

The forces acting on $ABCD$ are T_1, T_2, T_3, T_4 the pulls across AB, BC, CD, DA, respectively, and the resultant thrust, $p \cdot AB \cdot BC$, along OE.

Resolving along EF, we have
$$T_1 = T_3.$$
Similarly $\quad T_2 = T_4,$
and by Art. 132, $\quad T_1 = T_2.$

Resolving along OE, we have
$$(T_1 + T_2 + T_3 + T_4) \sin FOE = p \cdot AB \cdot BC,$$
$$\therefore 4T \cdot \frac{1}{2} \cdot \frac{AB}{r} = p \cdot AB \cdot BC,$$
$$\therefore 2\frac{T}{BC} = pr,$$
$$\therefore 2t = pr.$$

137. We have seen in Plateau's experiment with the oil in the water and alcohol (Art. 122), and have also deduced it from the hypotheses about surface-energy (Art. 124), that the separating surface between two fluids tends to become as small as possible. Now this is precisely what would happen if there were a stretched membrane separating the two fluids, as it would always tend to shrink; and it is easy to shew that if the tension throughout the membrane is constant and numerically equal to the surface-energy per unit area, the work done in increasing the area is equal to the total increase of surface-energy.

Let $PQRS$ be a rectangular element of the membrane, whose length PQ, a, and breadth PS, b, are both indefinitely small.

Let t be the tension throughout the membrane.

Let the membrane be first stretched so that QR moves parallel to itself to the position $Q'R'$ through the distance x.

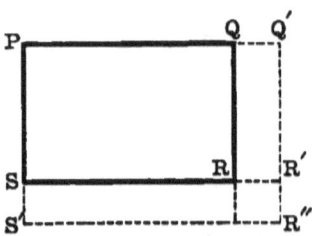

The work done in thus stretching the membrane
$$= t \cdot PS \cdot QQ' = tbx.$$

Now if the membrane be stretched so that SR' moves parallel to itself to the position $S'R''$ through the distance y,

the work done $= t \cdot PQ' \cdot SS' = t(a+x)y$.

∴ the total work done
$$= tbx + t(a+x)y$$
$$= t\{(a+x)(b+y) - ab\}$$
$$= t \times \text{increase in area.}$$

But the increase in the surface-energy, if S be the surface-energy per unit area, is $S \times$ increase in area,
$$= t \times \text{increase in area,}$$
$$\therefore t = S.$$

As any finite area may be divided into an infinite number of indefinitely small rectangles, and the theorem is true of each, it is true of the whole surface.

138. It follows from the above that we may investigate capillary phenomena, either as hitherto from the energy criterion of the equilibrium of a material system, taking

CAPILLARITY. 191

the surface-energy into account, or from the ordinary conditions of equilibrium assuming the existence of a uniform tension throughout each bounding surface, numerically equal to the surface-energy per unit area.

We may by the latter method obtain the result of Art. 125, by considering the equilibrium of an element of the line of intersection of the fluids. The result of Art. 126 may also be obtained by this method.

*139. Let us apply it to investigate the form of the *Capillary Curve*, which is a section of the cylindrical surface assumed by a liquid in contact with a vertical plate.

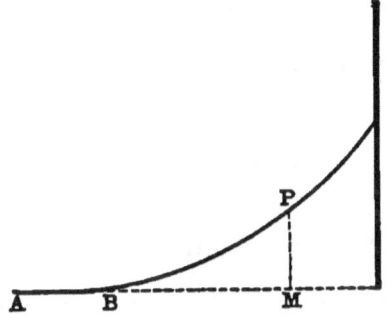

Let P be any point of the curve; draw PM vertically downwards to meet AB the horizontal surface of the liquid produced.

The pres. at M = internal pres. at $P + w \cdot PM$,

But pres. at M = atmospheric pressure = external pressure at P,

$\therefore$ external pres. liquid at P − internal pres. = $w \cdot PM$.

But (Art. 133), $t = pr = wr \cdot PM$,

where r is the radius of curvature at P, and t the surface tension.

Hence PM varies inversely as r.

The curve is therefore the same as the Lintearia (Art. 135).

140. The height a liquid rises in a capillary tube may be obtained as follows:

Let t be the surface-tension in the surface separating the liquid from the air. Let β be the angle at which this surface meets the tube. Let r be the internal radius of the tube, h the difference of level of the surfaces inside and outside, and w the weight of a unit volume of the liquid.

Consider the equilibrium of the column of liquid in the tube above the level of the surface outside.

The forces acting on it are

(1) its weight $\pi r^2 h w$ downwards,

(2) the tension at every point of the perimeter of the circle where the surface meets the tube. The resultant tension is clearly $2\pi r t \cos \beta$ upwards,

(3) the thrusts due to the pressure at the top and bottom of the column. These counterbalance as the pressure at both top and bottom is that of the atmosphere.

Hence (1) and (2) balance,

$$\therefore 2\pi r t \cos \beta = \pi r^2 h w,$$

$$\therefore 2 t \cos \beta = w h r,$$

which agrees with the result of Art. 129. In a similar manner we can obtain the rise of a liquid between two plates.

141. Soap Bubbles. The relation between the excess of internal over external pressure and the radius in an ordinary spherical soap bubble may be obtained from the relation $2t = pr$ (Art. 136) by regarding a soap bubble as a flexible membrane incapable of exerting a shearing stress. It should be observed that as t is here the total tension in the membrane, it is the sum of the tensions in the internal and external surfaces, i.e. it is twice the ordinary surface-tension. We may also obtain this relation by the principle of virtual work as follows:

Let S be the surface-energy per unit area in the bubble of radius r, p the excess of the internal pressure above the external.

If the radius be increased by an indefinitely small quantity x, the increase in the external and internal surfaces of the bubble together is $16\pi rx$, and therefore the increase in surface-energy is $16\pi rxS$. The work done by the pressure $= 4\pi r^2 px$, since each element of surface is pushed out through a distance x.

Therefore the equation of virtual work is
$$16\pi rxS = 4\pi r^2 px,$$
$$\therefore 4S = pr.$$

Ex. 1. For a soap bubble the surface tension is about 81 dynes per centimetre, find the difference between the pressure inside and outside a soap bubble whose radius is one centimetre. [Trin. Coll., 1883.]

Ex. 2. If the surface tension at different parts of a soap bubble were different, prove that it would not assume a spherical form, and shew how the variation of the tension would be indicated by its shape.
[St John's Coll., 1887.]

CAPILLARITY.

ILLUSTRATIVE EXAMPLE.

The surface of the liquid within a fine capillary tube is approximately spherical.

Let OA be the axis of the tube. Let $BPAP'B'$ be the section of the surface of the liquid in the tube made by a plane through OA.

Draw PM perpendicular to OA and produce it to meet the curve again in P'. Let PO, PT be the normal and tangent to the curve at P.

Let r be the radius of the tube, and let h be the height of A above the level of the liquid surface outside the tube.

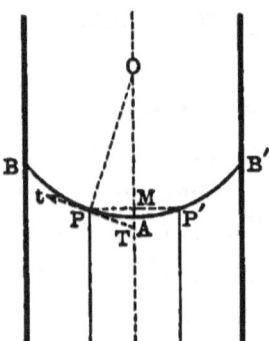

Let t be the tension throughout the liquid surface, w the weight of a unit volume of the liquid.

As in Art. 140, by considering the equilibrium of the cylindrical column of liquid of radius MP and length h, and resolving vertically, we obtain

$$\pi MP^2 \cdot hw = 2\pi t PM \cos PTO.$$

(The weight of the small portion of liquid above A is neglected, on the assumption that r is small compared with h.)

$$\therefore MP \cdot hw = 2t \cdot \frac{MP}{OP},$$

$$\therefore OP = 2t/hw, \text{ i.e. is constant.}$$

Hence it follows that the curve $BPAB'$ is a circle, of radius $2t/hw$.

If β be the angle at which the liquid surface meets the tube, we have (Art. 140)

$$rhw = 2t \cos \beta,$$

$\therefore$ the radius of the spherical liquid surface, $2t/hw$, is $r \sec \beta$.

EXAMPLES ON CHAPTER IX.

1. Prove that the surface of the liquid between two vertical parallel plates very near together is approximately a circular cylinder, and find its radius.

2. Two soap bubbles, blown from the same mixture, are allowed to coalesce into a single bubble: find an equation giving the radius of this bubble, and shew that this equation implies a reduction of the total surface. [St John's Coll., 1887.]

3. Explain how a drop of water can be in equilibrium in a fine vertical tube open at both ends.

4. A liquid film hangs in the form of a surface of revolution with its axis vertical. The upper boundary of the film is a circular wire held horizontally, the lower boundary is a heavy elastic thread, hanging freely in the form of a horizontal circle of radius r. The natural length of the thread is $2\pi a$, its modulus of elasticity is λ, and its weight is $2\pi a w$. The tension of the film is t. Prove that r satisfies the equation

$$(\lambda^2 - a^2 t^2) r^2 - 2\lambda^2 a r + (\lambda^2 + w^2 a^2) a^2 = 0.$$

[Trin. Coll., 1893.]

5. A soap bubble of radius a is blown inside another of radius b, when the atmospheric pressure is Π. When the atmospheric pressure changes to Π' the radii of the two bubbles are a' and b'. Prove that

$$\frac{b\Pi}{b'\Pi'} = \frac{b(a^2 - a'^2)(b'^3 - a'^3) + a^3 b'^3 - a'^3 b^3}{b'(a^2 - a'^2)(b^3 - a^3) + a^3 b'^3 - a'^3 b^3},$$

the temperature being supposed constant. [Peterhouse, 1892.]

6. If a frame of fine straight wire in the form of a tetrahedron be lowered into a solution of soap and water and drawn up again, there are found in certain cases plane films starting from the edges and meeting in a point. Shew that this is not a possible form of equilibrium for every tetrahedron, and that it is so if one face be an equilateral triangle and the others isosceles triangles whose vertical angles are each less than $\sec^{-1}(-3)$. [M. T., 1873.]

CAPILLARITY.

7. A spherical membrane contains air at a pressure ϖ, greater than the atmospheric pressure Π: if T be the tension in the membrane, r its radius, shew that the acceleration outwards of any point of the sphere is

$$\frac{4\pi}{Mr}\{\varpi_0 a^3 - \Pi r^3 - 2Tr^2\},$$

where M is the mass of the membrane, a its radius at some fixed time, and ϖ_0 the value of ϖ at that time.

If the membrane be a soap bubble whose equilibrium radius is a, shew that small variations about a in the value of r can take place, their period being

$$(\pi M)^{\frac{1}{2}}(4T + 3\Pi a)^{-\frac{1}{2}}.$$

[Clare Coll., 1891.]

8. Prove that of the energy required to blow a soap bubble $\frac{3}{5}$ is spent in forming the film and $\frac{2}{5}$ in compressing the air inside.

[St John's Coll., 1882.]

9. What will be the result if a capillary tube be bent below the level of the surface inside it and continued indefinitely in a horizontal direction? [St John's Coll., 1886.]

ANSWERS TO EXAMPLES.

CHAPTER II.

Page 10.

1. $888 \cdot \dot{8}$.   2. 13625, 81750, 40875, g being taken as 981.

Page 15.

1. $\cdot 345$ cu. ft.: 9280000.   2. $\cdot 0103$ ft.

Page 19.

1. 2 gallons.   2. 1.   3. $(vs + v's')/\sigma$.
5. $S_1 \pm \sqrt{(S_1{}^2 - S_1 S_2)}$.   6. 4·7 per cent.

CHAPTER III.

Page 28.

1. 36·864 ft. (v. Art. 16).   2. $\frac{1}{10}$ ft., $\frac{1}{10}\sqrt{5}$ secs.
3. 2486·5625.   5. 14·96 cu. inches.
6. 16·65 lbs. per sq. ft. approximately.
8. $\frac{1}{12800}$ ft., $\frac{3}{6400}$ lbs. per sq. inch.
9. At same level, $1\frac{21}{31}$ ft. above the horizontal tube.

PAGE 38.

1. For first bucket two-thirds of what it would be, if the bucket were at rest; for second, four-thirds.

2. A horizontal plane, (1) water above, (2) water below. (1) Pressure in water at depth z below the surface, $(g-f)\rho z$, at depth z below the mercury surface $(g-f)\{a\rho+z\sigma\}$. (2) At height z above water surface $(f-g)\rho z$, and $(f-g)\{a\rho+z\sigma\}$ at height z above mercury surface. a is thickness of water stratum, ρ, σ densities of water and mercury.

3. The inclination $=a$, that of the plane, and the pressure at distance z from the surface $=$ the atmospheric pressure $+g\rho z \cos a$, where ρ is the density.

PAGE 42.

1. Right circular cylinders having the axis as common axis. The pressure at distance r from the axis is $\dfrac{\omega^2\rho}{2}(a^2-r^2)$, where ω is the angular velocity, ρ the density, and a the radius of the free surface.

2. The volume $=\dfrac{\pi}{12}\cdot\dfrac{h^3\tan^2 a}{g}[4g-3h\omega^2\tan^2 a]$.

3. Pressure at distance r from axis $=$ pressure there when cylinder is at rest $+\frac{1}{2}\omega^2\rho(r^2-\frac{1}{2}a^2)$, where ρ is the density of mercury, and a the radius of the base.

PAGE 44.

1. $\frac{1}{4}\Pi a^2\sqrt{3}+\frac{1}{4}a^2 g\rho(\sqrt{3}\cdot c-a\cos\theta)$, where θ is the angle the line from the centre perpendicular to the base makes with the vertical drawn upwards. **2.** 6138 cubic feet.

3. $\frac{4}{5}(1+\sqrt{10})$ inches. **4.** Half the depth of the vertex.

5. If $ABCD$ is the parallelogram, AB being in the surface, the required line through A meets CD at a point distant $\frac{1}{4}CD$, from C.

6. $1:\sqrt{2}-1:\sqrt{3}-\sqrt{2}$. **7.** $\frac{11}{12}\sqrt{73}$ times the weight of the liquid. **8.** If h be the depth of the lowest side, the depths of the lines are

$$h\sqrt{\frac{1}{n}},\quad h\sqrt{\frac{2}{n}},\quad h\sqrt{\frac{3}{n}},\ \&c.\ \text{respectively.}$$

ANSWERS TO EXAMPLES. 199

CHAPTER IV.

PAGE 68.

1. If a, b be the lengths of the parallel sides, h the distance between them, the depth of the c.p. when (1) a is in the surface is $\dfrac{a+3b}{a+2b} \cdot \dfrac{h}{2}$, (2) one of the other two sides in the surface is
$$\frac{b^3 + b^2a + ba^2 + a^3}{2(b^2 + ba + a^2)} \sin \theta,$$
where θ is the inclination of the parallel sides to the vertical.

4. $16\frac{4}{9}$ inches below the oil surface.

CHAPTER V.

PAGE 84.

2. (1) $\tfrac{1}{12}g\rho a^3 \sqrt{2}$,   (2) $\tfrac{1}{12}g\rho a^2 \sqrt{(\tfrac{1}{2} 9a^2 + 27d^2 + 7ad\sqrt{6})}$, where ρ is the density of water.

5. If A be the area of the base, a the $\tfrac{1}{2}$ vertical angle, h the length of the axis, θ its inclination to the vertical, and ρ the density of the liquid, the thrust on the base $(P) = g\rho A \left\{ d + \dfrac{h \cos^2 a \sin^2 \theta}{\sin(\theta + a)\sin(\theta - a)} \right\}$, the thrust on the whole cone $(Q) = \tfrac{1}{3} Ah\rho \sin \theta$, and the thrust on the curved surface
$$= \sqrt{(P^2 + Q^2 - 2PQ \sin \theta)}.$$

9. g.   12. If h be the length, r the radius of the cylinder, θ the inclination of the axis to the vertical, d the depth of the centre, the required vertical thrust = wt. of vol. $\tfrac{1}{2}\pi rh \{r \sin^2 \theta + d(\sin \theta + 2\cos \theta)\}$ of the liquid.

PAGE 87.

1. 36·4 cu. cms. nearly: 7·5̇.   2. If ρ be the density of the cylinder, σ_1 and σ_2 the densities of the upper and lower fluids, the thrust on the upper end is to that on the lower :: $(\rho - \sigma_1)\sigma_1 : \rho\sigma_2 - \sigma_1^2$.

3. ·50065.   4. $15/\pi$ lbs. wt.

7. If W be the weight of the cone, W' that of the liquid displaced, and h be the length of the axis, the required force is $W - W'$ acting upwards at a point distant $\dfrac{(8W - 9W')h}{12(W - W')}$ from the axis.

11. The length of the heavier portion is one-quarter of the whole.
12. $1/\sqrt[3]{2}$.

PAGE 90.

2. $wv(1-\sigma)/g\sigma$.

PAGE 96.

2. (1) A hyperbola. (2) A point.   4. The sides horizontal and vertical, unstable: diagonals, vertical and horizontal, stable.

CHAPTER VI.

PAGE 118.

1. 8·8.   2. ·405 : 2·835.

PAGE 122.

1. 21.   2. 45/61.
3. The mass of silver is to that of gold :: 37380 : 37249.
4. 13·5.

PAGE 125.

1. 1·8 : ·9.   2. 4/3,: 2/3rds the distance of the 2nd graduation from the first.

PAGE 127.

1. 115 : 89 approximately.

CHAPTER VII.

PAGE 137.

1. 7 times the height of the water-barometer.   2. ·00013 cu. inches.

4. $\left(1+\dfrac{W}{\pi r^2 \Pi}\right)$ times the density of the atmosphere, where $W=$ the weight of the sphere, r its radius and Π is the pressure of the atmosphere.

5. Assuming that the height of the bell is small compared with the height of the water-barometer, the tension is increased.

6. If 1500 cms. be the depth of the *lowest* part of the bell, the height of water-barometer is 964·3 cms.; the pressure on the surface is 945014.

ANSWERS TO EXAMPLES.

Page 141.

2. The one whose temperature is the lower. $\Pi \left(\dfrac{1}{1+\dfrac{t}{273}} + \dfrac{1}{1+\dfrac{t'}{273}} \right)$.

4. $\dfrac{273+t}{v} \left(\dfrac{v_1 p_1}{273+t_1} + \dfrac{v_2 p_2}{273+t_2} \right)$.

5. $\dfrac{Vp}{76} \cdot \dfrac{273}{273+t} \times \cdot 1293187$.

6. $\dfrac{Vh\sigma}{3W} \cdot \dfrac{273000}{273+t}$.

CHAPTER VIII.

Page 162.

$\dfrac{Aln}{60B}$ ft. per second, if 1 ft. be the unit in terms of which A, B and l are given.

Page 166.

1. One-quarter of atmospheric density.

2. $\dfrac{l(l+2b)}{(l+b+c)^2} + \dfrac{b(b+c)(b+a)}{(l+b+c)^2(a+b+c)}$.

Page 169.

2. 22.

CHAPTER IX.

Page 184.

1. 1·2 inches.

Page 193.

1. 324 dynes per sq. cm.

INDEX.

Air-Pump, Hawksbee's, 164
——— Smeaton's, 165
——— Sprengel's, 167
——— Tate's, 166
Answers to Examples, 197
Archimedes' Theorem, 79

Balance, Hydrostatic, 120
——— Jolly's, 122
Balloon, The, 88
Barometer, The, 130
——— The Aneroid, 133
Barometric observation, Corrections to a, 131
Barometric observation, Determination of altitude by, 144
Boyle's Law, 134
Bramah's Press, 163
Bubbles, Soap, 192
Buoyancy, Surface of, 93

Capillary Curve, 191
Centre of pressure, Definition of, 63
——————— of any triangle, 68
Charles' Law, 138
Compressibility, Definition of, 14
Condenser, The, 168

Density, Definition of, 17
Diving-Bell, The, 137

Elasticity, Definition of, 14

Fire Engine, The, 163
Floatation, Surface of, 94
Floating Body, Conditions of equilibrium of, 86
Floating Body, Stability of, 91
Fluid, Definition of, 5
——— Definition of Perfect, 6

Gas, Definition of, 6
——————— Perfect, 136
Gauge, Barometer, 169
——— Condenser, 170
——— Siphon, 169

Homogeneous Atmosphere, Height of, 145
Hydraulic Press, The, 163
Hydrometer, The Common, 123
——————— Hare's, 119
——————— Nicholson's, 125

Impulsive pressure, 45
Inverted U-tube, Determination of Specific Gravity by, 119

Lintearia, The, 188
Liquid, Definition of, 7

Manometer, Compressed-air, 170

INDEX.

Membranes, Flexible, 184
Metacentre, Determination of, 96

Perfect Fluid, Definition of, 6
——— Gas, ——————— 136
Plateau's Experiment, 175
Pressure at a point, Definition of, 9
——— Impulsive, 45
——— Surfaces of Equal, Definition, 30
Pressure, Transmissibility of liquid, 15
Pump, The Common, 159
——— Force, 161
——— Lift, 160
——— Plunger Pole, 161

Quincke's Experiments, 176

Siphon, The, 153
Soap Bubbles, 192

Specific Gravity Bottle, 116
——————— by U-tube, Determination of, 118
Specific Gravity by inverted U-tube, Determination of, 119
Specific Gravity of mixtures, 18
Surface-Energy, 177
——— -Tension, 185
Syringe, The, 157

Temperature, Absolute, 139
Thrust on plane area, 43
Torricelli's Theorem, 154
Transmissibility of liquid, 15

U-tube, Determination of Specific Gravity by, 118
U-tube inverted, Determination of Specific Gravity by, 119

Valves, 158

www.ingramcontent.com/pod-product-compliance
Lightning Source LLC
Chambersburg PA
CBHW020827230426
43666CB00007B/1130